GETTING PRICES RIGHT

The Scope and Limits of Agricultural Price Policy

GETTING PRICES RIGHT

The Scope and Limits of Agricultural Price Policy

C. PETER TIMMER

Cornell University Press *Ithaca and London*

First published 1986 by Cornell University Press.

International Standard Book Number (cloth) 0-8014-1960-3
International Standard Book Number (paper) 0-8014-9441-9
Library of Congress Catalog Card Number 86-47649
Librarians: Library of Congress cataloging information appears on the last page of the book.

The paper in this book is acid-free and meets the guidelines for permanence and durability of the Committee on Production Guidelines for Book Longevity of the Council on Library Resources.

Contents

Preface

As an analytical exercise, agricultural price policy is a fashionable topic, which means that interest comes and goes. During a seminar given at Harvard University in 1980 on "the scope and limits of agricultural price policy in Sri Lanka," which served to give early expression to some of the ideas in this book, a look of puzzlement gradually spread over the face of Lester Gordon, then the director of the Harvard Institute for International Development. "I thought these were the issues our advisers were grappling with in Pakistan in 1960. Haven't we learned anything in two decades?"

Lester's remark provoked an impassioned response. After a flurry of interest in the early 1960s, the economics profession all but abandoned agricultural price policy as an academic pursuit. Although practitioners continued to face these pricing issues year in and year out, in fact little progress was made in understanding what the "right" prices were in any given circumstance. Indeed, it was my sense then that much of the academic work on price policy was wrongheaded—worrying excessively about optimal shadow prices to be used by planning agencies that could not keep typewriters working efficiently, much less an entire economy. My message was that markets would have to be used much more openly and intensively as vehicles for implementation of government objectives. The income distribution consequences of such a market approach, especially the nutritional consequences when food price policy is considered, would then be the main focus of analysis and of government programs. As my colleagues and I wrote in 1983 in *Food Policy Analysis*:

> Most countries start from a food price policy environment that uses food imports and budget subsidies for across-the-board consumer protection, while a host of production-oriented government projects attempt to increase food output. One of the primary lessons that comes from understanding the complexity of hunger and its connections to the food system suggests that such a price policy/project orientation is backwards. Governments can more effectively meet the full range of food policy objectives by using price policy, not to keep food prices low for consumers, but as part of the incentive package that induces greater food production from millions of small farmers. Programs and projects can then provide targeted food subsidies to protect the very poor until they find jobs and higher incomes that result from the new policy environment. (p. 14)[1]

Despite the complaints of one reviewer, F. C. Clift, that the book should have been called *Food Price Policy Analysis* because of the pervasive discussion of the role of prices in a food system, the book did not attempt to draw together the lessons for analysts and practitioners on how to think systematically about the impact and desirability of a change in agricultural price policy. That is the purpose of this slim volume. I attempt here to follow the advice of the reviewer of an early and partial draft produced in 1983 for the World Bank: the essay should stand "by itself as a single-source primer on farm price policy interventions and their many powerful consequences for pro-

[1]Full citations appear in the References section.

ducers (large and small), consumers (rich and poor), budgets, rates of growth, inflation, and trade" and incorporate what are likely to be the even more important dynamic and political economy ramifications.

An extended essay on the scope and limits of agricultural price policy must obviously draw heavily on a broad base of my own work and especially that of a multitude of others. The references provide a glimpse of the diversity of authors and topics that intersect this issue, but additional recognition must be given to the colleagues, institutions, and issues that have prodded and molded my thinking. What is now a decade and a half of involvement with Indonesian food policy is foremost on all three counts. My debts to BULOG, the Indonesian Food Logistics Agency, to BAPPENAS, the National Planning Agency, and to the other ministries and organizations that have patiently permitted me to learn about the process of economic development are boundless. Individuals too numerous to name sat on the other side of the table answering questions, but Dr. Saleh Afiff, my first mentor in Indonesia, who was then agricultural bureau chief in BAPPENAS and is now minister for administrative reform, General Bustanil Arifin, minister of cooperatives and chairman of BULOG, and General Sukriya Atmaja, vice chairman of BULOG, must be singled out as sources of support, wisdom, and encouragement.

Closer to home (but often in Indonesia), Wally Falcon and Scott Pearson at Stanford University have served as a more or less continuous sounding board for the ideas worked out in this essay. Although I have written the words, they surely share some of the credit (and, I must say, some of the blame as well) for the attempt to integrate the analysis from micro to macro and from short run to long run. Others who have provided helpful comments are Michael R. Reich, Peter Hopcraft, Isabelle Tsakok, and participants at a variety of seminars at Harvard, the World Bank, Stanford, and Indonesia.

These ideas have also been worked out in various teaching environments, including my course at Harvard on the world food system, the short courses on price policy sponsored by the World Bank, and a series of courses at BULOG given jointly by Falcon, Pearson, and me, as well as in several sessions at the Economic Development Institute of the World Bank in which I condensed all these ideas into a morning or an afternoon. When Price Gittinger of the Economic Development Institute asked if I could write up one of those presentations into a little paper, my response was just the opposite. The initial 98-page manuscript has blossomed into this longer effort.

Usually my editor keeps such tendencies firmly under control. My wife Carol has by now, however, become so much my alter ego on this topic that she may have lost her objectivity about what goes into the wastebasket. Her role this time has been to push for fuller explanations from me and to improve the clarity of expression herself. To the extent that the reader finds the material clear and comprehensible, we both have Carol to thank.

C. PETER TIMMER

Waban, Massachusetts

GETTING PRICES RIGHT

The Scope and Limits of Agricultural Price Policy

1

Introduction to Agricultural Price Policy

What is the "right" price for an agricultural commodity? Economists have an easy answer to the question, but only in a world of perfect information, with competitive markets, without other government interventions into the economy, and without political concerns for the impact on income distribution. In such a world, any deviation of the domestic price from the international border price of a commodity, as either an import or export, reduces total economic welfare in the country because of dead-weight efficiency losses in production and consumption.

Despite the many provisos and restrictions noted for this "answer" to be completely correct, the border price paradigm as a guide to domestic price policy remains quite powerful. In practical terms, the border price represents a society's short-run opportunity costs with respect to changed consumption or production of the commodity. Within the normative framework of economic theory, welfare is maximized by using border prices to determine domestic prices. Consequently, understanding the logic of the border price paradigm is the essential first step in addressing any further policy concerns over the implementation and impact of changes in domestic prices for food and agricultural commodities.

Understanding the border price paradigm is the starting point for analysis of agricultural price policy—not the conclusion. It is essential that all participants in a policy analysis and debate start at the same point, with a common framework and set of assumptions. From that initial point, representatives of various interest groups, ministries, and methodological perspectives can each build toward a more complex understanding of the pricing issues and formulate their concerns over the impact of a pricing change within an agreed paradigm.

These complexities and concerns stem from three quite different problems with the border price paradigm. First, most economists recognize that the underlying assumptions do not hold precisely in even the best of circumstances and that they are wildly violated in many developing economies, where knowledge is highly imperfect, competition and markets are badly fragmented, and extensive government involvement makes piecemeal analysis inappropriate. Consequently, the analysis must be robust enough and based sufficiently in empirical understanding of the actual situation that challenges to the theoretical model can be met with relevant adaptations and facts.

Second, the political concerns for income distribution simply cannot be assumed away, whether they reflect the powerful vested interests of urban workers and the military or a genuine concern for the nutritional well-being of the poor who depend on access to cheap food for their day-to-day survival. Changes in prices of important agricultural commodities inevitably carry visible consequences for the welfare of the poor. This concern is a reason not for keeping all prices constant for all time but rather for understanding the impact of price changes on welfare and their political implications. Since political impact is almost always felt much more rapidly than full economic impact, the analysis must focus specifically on the time dimensions of both and illuminate possible side interventions that will make desirable economic changes politically feasible.

A third level of complexity and concern comes with problems of implementing price policies. Operational ministries are concerned about administrative procedures for implementation of a price policy and the impact of policy changes on the marketing sector in which prices are formed and communicated. They are also concerned to know what the international price will be for the planning horizon of the price policy. International commodity markets are notoriously unstable; a domestic floor price policy announced before planting season, for example, may well reflect the opportunity cost for imports at the time of the announcement, but the floor price could be too low or too high by the time of harvest. Instability in international commodity prices is thus a serious issue for the design of domestic price policy in the short run, whereas long-run structural changes in the world food system affect investment decisions about agricultural research, rural infrastructure, and sectoral priorities.

At the same time, the international price, whatever it is, must be converted into a domestic currency equivalent. What is the appropriate exchange rate for such a conversion? In principle, economists suggest that a "shadow price" that represents the opportunity cost of foreign exchange to the country should be used. Much of the literature on project appraisal and national planning is devoted to finding shadow prices (whose use by planning agencies would maximize total welfare for the society). In practice, the shadow rate of foreign exchange is not known or agreed to by all policy makers, which makes it awkward to use in the design of price policy. Consequently, international prices are usually converted at current exchange rates, with all the biases implied by those rates. An important part of the discussion of *agricultural* price policy is thus about *macro* prices, especially the foreign exchange rate.

The analysis of agricultural price policy has two concerns: understanding the influence of current prices on the structure of production and the distribution of welfare; and tracing the dynamic impact of changes in those prices. The standard partial-equilibrium analysis using border prices as the measure of domestic opportunity cost addresses these issues at one level by measuring gains and losses in producer and consumer welfare, transfers to and from the budget, changes in foreign exchange earnings and expenditures, and dead-weight efficiency losses. But these welfare effects are at an aggregate level only. They do not distinguish among the consequences for low-income and high-income consumers, for example, or among the various producers who receive or pay the transfers implied by domestic prices set at a level different from world prices.

Determining the welfare consequences of price policy requires that standard economic welfare analysis be disaggregated at least by income class, and sometimes by region, ethnic group, or age and gender. Especially when the price policy is for a major food commodity, such disaggregated analysis should search for any important changes in nutritional status likely to be caused by significant changes in pricing.

The initial framework for the welfare analysis of changes in price policy is a static and partial equilibrium one. The dynamic and general-equilibrium consequences of price changes are extremely hard to measure or predict, since no economy is understood well enough for these effects to be modeled except in the roughest, intuitive way. But this does not mean that the consequences are unimportant or that price policy analysts could or should ignore them. Incorporating the impact of changed food and agricultural prices on other markets, on the structure of investment, and on long-run patterns and distribution of economic growth is also an important analytical task, no matter how rough the answers may turn out to be. These issues also take the analyst of price policy into the macro policy debate, this time via budgetary, fiscal, and monetary issues in addition to the debate over macro price policy.

This essay attempts to illuminate and connect these diverse aspects of appropriate price policy in the following sequence: (1) the implementation issues and impact on the domestic marketing sector, with special consideration for the relative roles of public and private marketing agencies as a function of price policy; (2) the nature of the world market price; (3) the disaggregated effects of price policy on producers and consumers; (4) the short-run macro effects on budgetary, fiscal, and monetary policy; (5) the impact of macro prices, especially the foreign exchange rate, on agricultural price policy; (6) the spillover effects of price policy for one commodity market on other commodity and factor markets, on the agricultural sector as a whole, and ultimately on the entire economy; and (7) the dynamic effects through micro decision making and the macro economy on employment, the composition and size of investment, and the structure and rate of economic growth.

Table 1.1 illustrates these pieces and relates them to the starting point—the border price paradigm for economic welfare analysis of price policy interventions. As in all economies, each piece is related to all the others, and starting the analysis anywhere should lead it everywhere. There are pedagogical advantages, however, to beginning with the basic and familiar analytical framework of welfare analysis and then extending it from the implementation issues—impact on the domestic marketing sector, world market price instability, and disaggregation—to the less easily understood issues—short-run macro effects, the dynamic impact on the distribution and rate of growth, and the changing role of agricultural prices in a country's political economy as it undergoes the process of structural change.

There is a danger that analysts will become enamored of the elegance and quantifiable tractability of the partial-equilibrium

Table 1.1. Components of price policy analysis

Framework for beginning the analysis		Components of a complete agricultural price policy analysis
Partial equilibrium	⟶	General equilibrium multi-market agricultural sector
Comparative statics	⟶	Dynamic economic political
Market aggregates	⟶	Disaggregated
Prevailing border price	⟶	Long-run opportunity costs
	Additional analysis needed	
Marketing and implementation Understanding and using international markets large county effects structural changes instability Impact on producers and consumers by income class Impact of budgetary, fiscal, and monetary policy; feedback mechanisms to inflation		Role of macro prices foreign exchange rate interest rates wage rates Macroeconomic consequences Dynamics of price policy investment and technology income distribution and savings rate and patterns of economic growth creation of vested interests

model of price policy. To a certain extent this is a healthy reaction; an intimate understanding of the entire neoclassical pricing paradigm provides several launching platforms into the complexities and messiness of imperfect information and incomplete markets, unstable border prices, distorted foreign exchange rates, political impact, and dynamic ramifications. Contemplating all of these problems without some basic framework of analysis leads to a paralysis of the planning process, which explains why agricultural price policy in so many countries appears to be so mindless or contrary to the best interests of their citizens. Without an analytical framework, political ideology easily dominates the decision-making process of policy makers. Any price policy can be justified on some political basis. The problem for analysts of price policy is to understand this potential and to have a robust framework for working out the economic and political consequences in both the short run and long run. Knowing the implications of the border price paradigm prepares the analyst to deal with the more complicated and more important issues that are certain to arise.

2

Price Policy Analysis: The Partial-Equilibrium Framework

All governments must intervene in their economies to raise the budgetary resources needed to carry out even the minimum tasks of government: provision of police protection and roads, and other inherently public activities. Taxation of an economic base in thus the starting point for a government's sheer existence. Wherever that economic base is primarily agricultural, direct or indirect mechanisms for taxing agricultural incomes are likely to be essential components of government policy. Given a lack of political desire or bureaucratic capacity to tax factor incomes in agriculture directly, through land, profit, or personal income taxes, the widespread use of commodity taxes as a basic source of government revenue is inevitable. In poor countries price interventions into agricultural commodity markets are almost essential for the functioning of government. This "need to tax" exists even when the taxes would otherwise be considered as distorting market signals to producers and consumers, and some balance between the need for revenue and the magnitude and impact of the distortions must be struck. Price policy analysis can help identify the trade-offs in given circumstances and can even point the way to more efficient sources of revenue. But the analysis cannot eliminate the dilemma and the need for government price interventions.

Even apart from the need to raise tax revenues, governments intervene into agricultural price formation for a purpose. Sometimes that purpose is narrowly self-serving to those in power and their friends; sometimes it is meant to serve the broader objectives a society holds for its food and agricultural sector. Policy analysis is ultimately normative. It is designed to illuminate the consequences of a movement of a food system in some desired direction. What is desired obviously varies from society to society and over time. With appropriate weights attached to reflect this variance, the following general objectives for the food sector are representative of the goals pursued by most societies. The policies are meant to

contribute to efficient economic growth;
improve income distribution;
ensure a nutritional floor for all citizens;
guarantee food security for the country;
maintain political stability, which is probably assured if the first four are achieved.

Some of these objectives would be achieved, at least in part, in the absence of direct policy intervention by the government, but most governments actively seek to design and implement policies that are intended to improve this "hands-off" performance from the perspective of key interest groups in the society. Nearly any government intervention has at least indirect effects on food prices facing consumers or crop prices facing farmers and influences agricultural incentives in subtle and roundabout ways (see Timmer, Falcon, and Pearson). But the reverse causation is also important and is not developed in that volume. Agricultural prices have subtle and dynamic effects on the entire economy in addition to their direct and more immediate impact on the agricultural sector itself. These issues are pursued later in this volume, but the question now is more specific: What are the direct economic consequences of a government price intervention that causes domestic producer or consumer prices, or both, to deviate from their "free-market" level?

The Market Paradigm

The free-market level of prices is subject to the workings of the entire economy along with the full range of government policies and programs, and these factors are reflected in the national supply and demand curves for the commodity in question. Such supply and demand curves are the starting point for the analysis, but both the slopes and positions of these curves are to some extent determined by an earlier history of prices and price policies. Likewise, the supply and demand curves five or ten years in the future will be similarly influenced by the price policies currently under analysis, although the initial stages of analysis take little cognizance of this fact. Empirically, these curves can be represented by initial quantities and estimated supply and demand elasticities applicable to small changes in prices. Very few commodity systems in any country are well enough understood to draw either supply or demand curves over an extended range.

Figure 2.1 shows a standard result. The supply curve intersects the demand curve at an equilibrium price P_0 at which quantities supplied and demanded are equal to each other at level Q_0. As a market outcome, equality of supply and demand is a simple empirical result. No ethical or welfare connotations are involved unless strict conditions hold for the processes that generate these curves. In particular, the supply curve must re-

Figure 2.1. Supply and demand framework for price formation

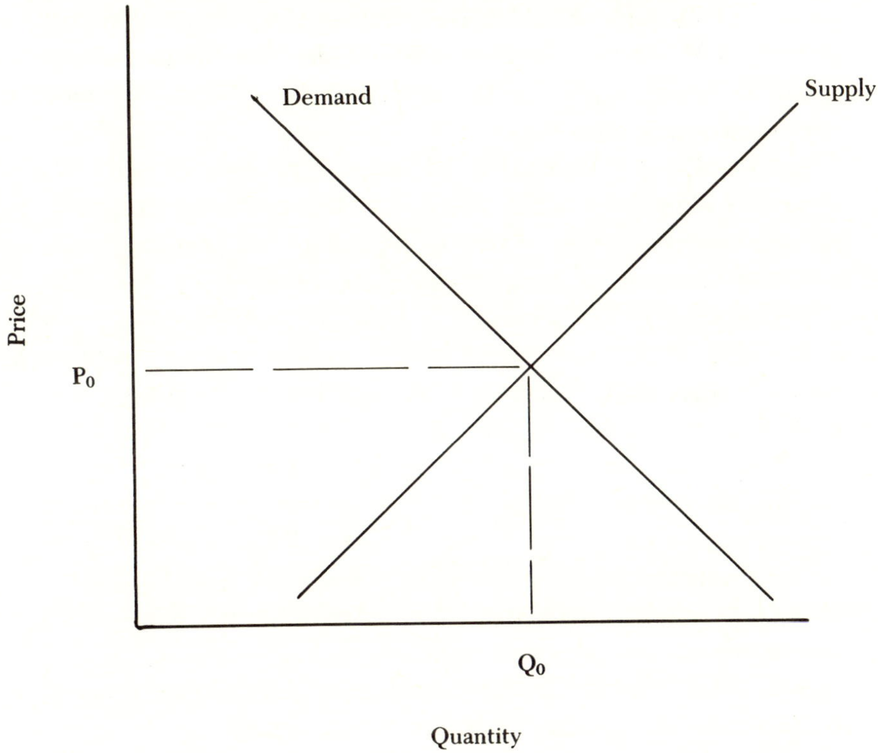

flect the social marginal costs of producing each level of output, whereas the demand curve must reflect the marginal value of gains to consumers at each level of consumption. If the empirical supply and demand curves for all commodities meet these normative conditions, market equilibrium leads to the equality of social marginal costs and social marginal benefits, and to Pareto optimality for the existing income distribution—no one can be made better off by reallocating society's resources in either production or consumption without simultaneously making someone worse off. A different distribution of income, of course, might lead to a quite different allocation of resources with the same optimality conditions holding. In conducting price analysis, one should know whether the estimated supply and demand curves that represent empirical responses correspond at least roughly to the normative supply and demand curves of economic theory.

A second factor implicit in the market-clearing equilibrium shown in Figure 2.1 is the isolation of this particular commodity market from other markets, especially international markets, for the same or closely related commodities (in the sense that Thai rice is closely related to Japanese or Indonesian rice). This isolation can come about for a variety of reasons, many of which involve no government intervention at all. High transportation costs or the physical absence of markets are the most common nonpolicy reasons, but a policy of trade autarky for the commodity would produce the same result. The lowering of transportation costs, the creation of more substantial and regular international markets through technical change for important crops, or the realization that autarky has serious welfare consequences can alter the isolation of the commodity market shown in the figure. This opens the possibility of international trade in the commodity, and the consequences can be examined using Figure 2.2.

Figure 2.2 illustrates a situation in which lower transportation costs or a newly found abundance of grain in world markets, for example, make it possible for profit-motivated private commodity traders to offer the commodity (in unlimited quantities) at the border of the country at price P_w, the "world" price. The development of railroads and steamships, hybrid corn, and fertilizer-responsive varieties of rice and wheat each has contributed in the last one hundred years to such opportunities for trade. In the absence of government efforts to stop it, such trade occurs more or less on its own volition as traders seek opportunities to make purchases and sales. The figure shows an opportunity for international grain dealers to make a

Figure 2.2. Price formation with international trade

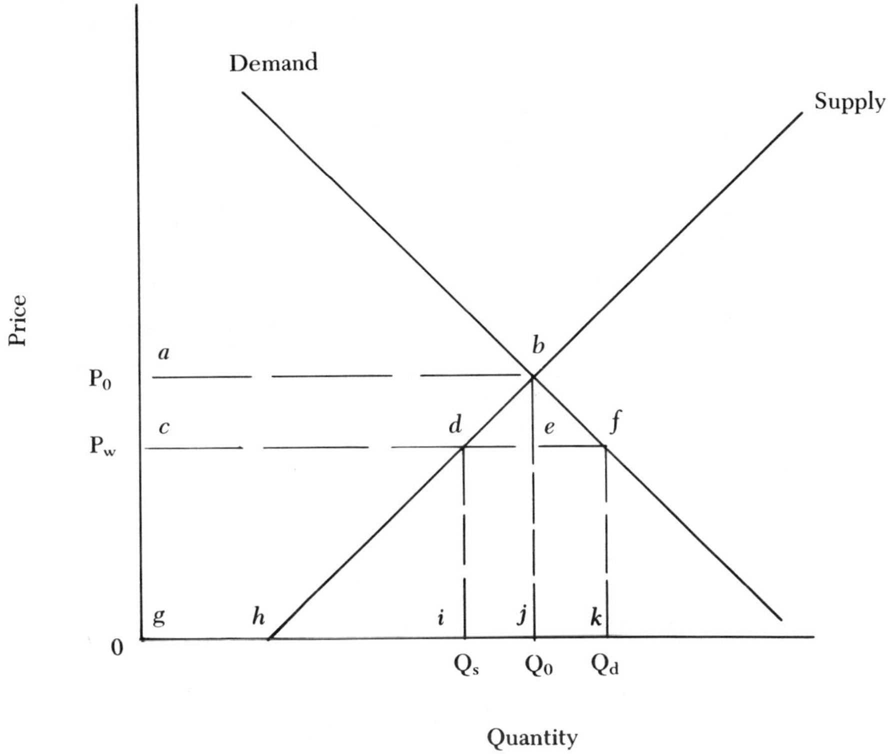

sale: the world price is below the domestic market-clearing price under autarky. The outcome is now quite different from the earlier result without trade. If rice, for example, is available on a delivered basis from world markets at price P_w, consumers will not pay more than P_w, and so farmers cannot receive more than P_w. The equilibrium quantities produced domestically and consumed are no longer equal, because demand increases to Q_d while local farmers reduce their supply under the pressure of world competition to Q_s. The gap, $Q_d - Q_s$, must be supplied by imports from the world market at a price of P_w.

Several important assumptions are hidden in this apparently

simple result. The world price P_w is quoted in terms of the domestic currency—rupees, pesos, or the like—in order to be comparable to the domestic prices shown in Figure 2.1. But the actual world market price is usually quoted in dollars or another standard international currency, and a conversion at some exchange rate is required. The rate that is used clearly has a significant effect on how P_w compares with P_0. For several countries, the size of the import gap $Q_d - Q_s$ that arises from a given P_w actually influences the equilibrium level of P_w itself through an impact on the country's own foreign exchange rate. The imports needed to fill the gap might also influence P_w in world markets. This is a major topic of Chapter 4. Similarly, even the dollar quotation of P_w can be known with certainty only for a particular delivery date and set of conditions. The uncertainty in this price for longer-term planning also is treated in Chapter 4.

More important in the short run, several adjustments take place in the domestic commodity market illustrated in Figure 2.2. An understanding of these adjustments caused by international trade can help determine the impact of price policy, for two reasons. First, the opportunity to trade has changed the price facing domestic producers and consumers. Whatever the reason for the price change, this is an opportunity to see the impact of price changes on these decision makers and, perhaps more broadly, on other markets and the entire economy. Second, most price policies are implemented by interventions at the border—by taxing or subsidizing international trade in the commodity. It is quite difficult to implement a price policy in the absence of international trade in the commodity concerned; the mechanisms to intervene are much more complicated and require direct purchases and sales by a government-controlled marketing agent. Trade interventions, by contrast, can usually be implemented quite simply by the customs service, a government trading company, or both.

Under autarky, domestic production equals domestic consumption, and no foreign exchange is needed or generated by the market outcome for the commodity. Farmers may need foreign exchange indirectly, of course, to pay for their fertilizer or tractors, or even for the imported radios or motorcycles which may be a part of their consumption bundle made possible by the high incomes earned by producing the commodity at price P_0. Similarly, nonfarm consumers may have less need for foreign exchange to pay for such imported goods because much of their income is devoted to purchasing the domestically produced commodity at the relatively high market-clearing price P_0. But no foreign exchange is needed directly to import the commodity.

This situation changes when rice, for example, is imported. In Figure 2.2 the domestic price falls from P_0 to P_w, and the import gap $Q_d - Q_s$ requires the direct expenditure of foreign exchange corresponding to the area of rectangle *dfki*, or $P_w (Q_d - Q_s)$. Where will this foreign exchange come from? If these are simply empirical supply and demand curves, the answer is "hard to tell," and maybe there will be a foreign exchange constraint on such a free-trade policy. But if these supply and demand curves roughly reflect underlying social opportunity costs and benefits and if the price change is relatively small, the process of market adjustment *automatically* frees up the necessary foreign exchange in a manner discussed below.

It is important to realize that this process of market adjustment is not shown in Figure 2.2, which reflects only the single commodity market, for rice, under investigation. The market for foreign exchange operates elsewhere, linked to the rice market (and others) but functioning separately. If the rice market is important (by a variety of criteria to be discussed later in this chapter), the impact on the foreign exchange market of moving rice prices from P_0 to P_w may be significant. Even if all markets are working perfectly, this impact may be of policy interest; where some of the underlying assumptions of the competitive model are substantially wrong, separate analysis of the foreign exchange market will be needed before a full assessment can be made of the effects of the initial price change on the rice market itself. For example, if the foreign exchange required to pay the import bill *dfki* raises the equilibrium foreign exchange rate, the border price P_w is raised as well. This price change in turn raises domestic production above Q_s, lowers domestic demand below Q_d, and reduces the import gap and number of dollars needed to pay the import bill. (The amount of domestic currency might actually increase, however.)

Imports of rice at price P_w cause other important adjustments in Figure 2.2. Consumers are better off for two reasons: they can buy their original quantity of rice Q_0 at the lower price P_w, and they can increase the quantity consumed to Q_d. Their net welfare gain is the area *abfc*, the increase in consumer surplus. This is the amount of money consumers would have been willing to pay to consume the additional quantities between Q_0 and Q_d but do not need to pay because they can buy all the rice they want at P_w.

This gain in consumer welfare comes partially at the expense of lower incomes for domestic producers. Without trade, total producer revenues (equal to total consumer expenditures) were *abjg*; if the supply curve measures the relevant marginal costs, total producer costs are the area under the supply curve, *bjh*, and producer surplus, or income, is equal to the difference, *abhg*. When the price falls to P_w, producer incomes fall by the amount in area *abdc*. This is a quite quantifiable loss in producer welfare and is transferred, via the lower prices in the rice market, to consumers.

It is apparent that consumers have gained more than producers have lost, and the government has done nothing more than allow free trade in rice. Even if consumers could in some way precisely (and without transaction cost) reimburse producers for their losses, consumers would be net gainers by area *bfd*. This triangle represents the "gains from trade," gains generated by the increased efficiency in the allocation of resources made possible through the opportunity to trade. Again, it is important to remember that other potentially significant adjustments must occur for this efficiency—and the welfare gains that come from it—to be realized. In particular, resources must be free to flow from one market to another to equilibrate opportunity costs and returns in all activities.

The potential resource transfers can be identified in Figure 2.3. Assume that producers and consumers have similar expenditure patterns, so that the transfer of income *abdc* from producers to consumers does not require any switch in resources. (Such a switch is likely, of course, because rural and urban households tend to have different expenditure patterns, but it unnecessarily complicates the point here.) Producers have withdrawn resources from growing rice as production was cut back from Q_0 to Q_s; these saved resources are equal to area *bjid*. At one extreme, these resources might be imported fertilizer, and so the resource savings directly generate part of the foreign exchange needed to import the rice that would otherwise be grown domestically. At the other extreme, the resources might be labor or land that grows nothing but rice (at least in the short run) and hence are simply wasted under the lower price. If their opportunity cost to farmers is actually zero, all available resources will always be committed, and the supply curve should be much steeper or even vertical, which indicates that short-run supply elasticities are nearly zero. Factors other than fixed resources can also cause low short-run supply elasticities (see Streeten).

Consumers represented in Figure 2.3 have also freed up resources by switching some of their purchasing power away from other commodities toward rice because of the stimulus of lower rice prices. These resources are presented by area *efkj*. If these

Figure 2.3. Transfers of economic resources due to international trade

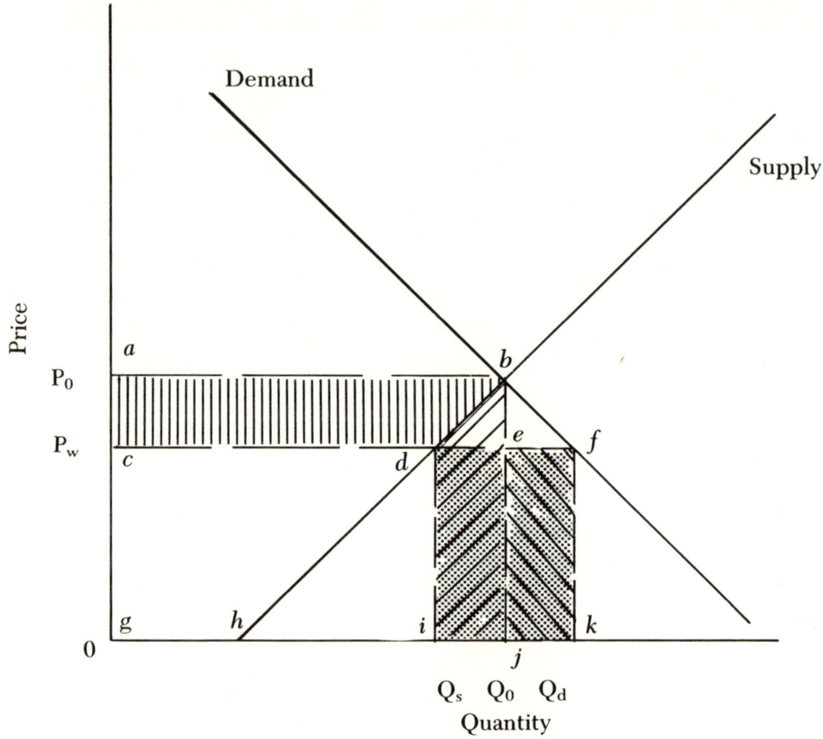

Producer income transferred to consumers (*abdc*); assumed to be spent on same goods and services with no reallocation of production needed

Resources no longer used to grow rice (*bjid*) and available to grow other crops; could be imported (fertilizer) or domestic (land, labor)

Additional consumer resources devoted to rice consumption that are diverted from other goods and services, freeing them for other uses (*efkj*)

Domestic resources needed to procure imported rice: assumes foreign exchange available at a rate reflecting its opportunity cost to the economy (*dfki*)

consumer expenditures had originally been spent on imported textiles, for example, and are now simply switched to imported rice, the foreign exchange required is generated directly, and no adjustments need to be made by other markets. If producer costs and consumer expenditures are not so neatly cast in foreign exchange terms, other markets must adjust. But the resources exist for those adjustments to take place. The land and labor involved in area *bjid* might grow export crops, for example. The equipment and labor involved in producing domestic goods purchased with the incomes represented in *efkj* might switch to producing for export markets when domestic market demand drops as consumers buy more rice. If domestic markets are competitive and flexible and if export demand exists for domestically produced goods, the resources are generated in the partial-equilibrium framework shown in Figure 2.3 needed to lead to a fully consistent result, even though there may be important spillover effects in these other markets that are not identified directly within the framework.

Price Policy Interventions

A government contemplating the free-trade results for a food commodity in Figures 2.2 and 2.3 has three direct options and a wide variety of indirect interventions available. First, it can simply accept the free-trade results, with their consequences for both income distribution and efficiency. Second, even the free-trade price may seem too high for consumers, and policy can then be directed at lowering food prices, perhaps to improve nutritional status and urban income distribution in the short run. A third alternative is to protect domestic producers from foreign competition by raising farm-level prices for the commodity. This measure would provide incentives for greater output and might improve income distribution between rural and urban areas. If the size of the import gap leaves policy makers insecure over the source of the country's food supplies, higher domestic food prices might provide a sense of greater food security at the macro level (while leaving poor households with reduced food security). Imports would be less, perhaps zero, and direct foreign exchange needs would be reduced.

It is also possible to direct policy at *both* groups by lowering consumer prices and raising farm prices. This requires a narrower (or even negative) marketing margin between farmers and consumers. The costs and implementation problems of such a dual price policy can be quite substantial. The relationships between marketing costs, marketing agents, and price policy are dealt with in Chapter 3. An extended treatment of dual price policies is available in Timmer, Falcon, and Pearson (Chap. 4) and in several research reports of the International Food Policy Research Institute. Attention is restricted here to policies that either raise or lower the domestic price for both producers and consumers.

Subsidies to Lower Food and Agricultural Prices

A government can attempt to lower a commodity's price in several ways, including coercion, legal price controls, and exhortation. The only way to lower prices for all consumers on a reliable basis for more than brief periods, however, is to increase the supplies available in the market relative to the demand. If demand is not to be lowered (by rationing, for example), supplies must be increased, and barring a rapid and effective shift outward in the domestic supply curve, imports would be needed to provide these supplies. Because the price of imports converted at the current exchange rate is higher than the desired market price (hence the government's decision to lower prices), subsidies from government revenues must be used to close the gap between the world price and the

Figure 2.4. Price formation with a subsidy on imports

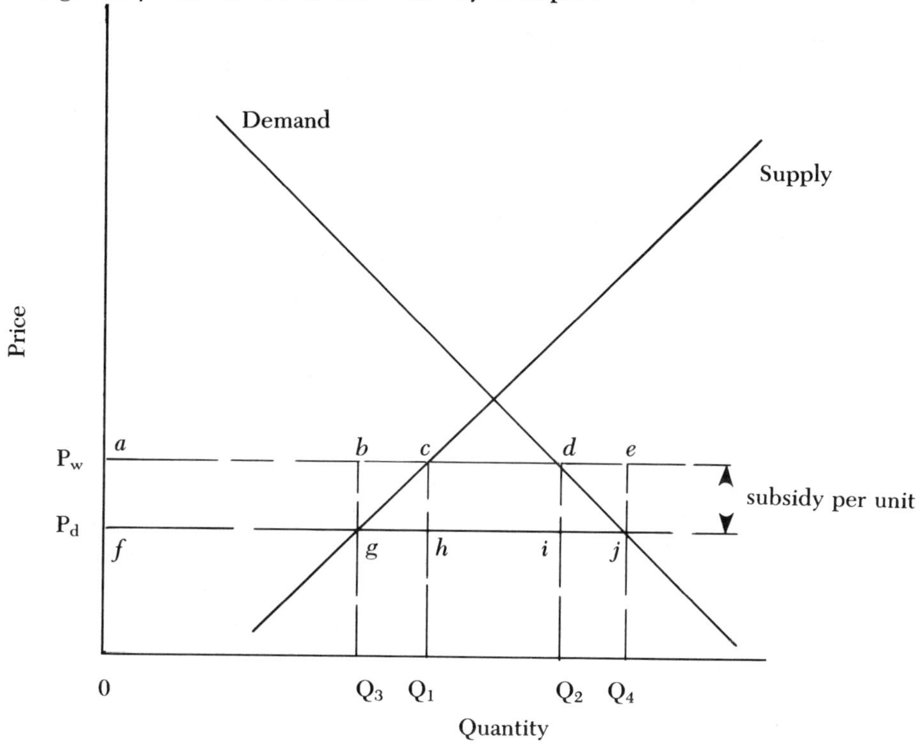

desired domestic price. Figure 2.4 presents a schematic illustration of the changes involved.

The supply and demand schedules in Figure 2.4 are identical to those shown in Figures 2.1, 2.2, and 2.3. These curves reflect the initial status of technology and productivity on the supply side (including the costs of inputs and alternative commodities that farmers might grow instead of the commodity in question). On the demand side, the curves reflect the tastes of consumers, the level and distribution of incomes, and the prices of alternative consumer goods (including not only other food items but also nonfood goods and services).

It is because all of these factors influence the location of the supply and demand curves that they represent only a partial-equilibrium reaction to a change in the price of the commodity being considered. When the commodity is important to the entire economy, as a staple food grain or major export crop inevitably is to a developing country, these partial-equilibrium adjustments provide only an initial glimpse of the full adjustments that are likely to occur when the price changes. Other, more complicated analytical techniques must then be used to understand the full effects of price policy. The nature of these effects, and the types of models available to analyze them, are discussed in Chapters 6 and 7.

Spillover effects. The simple model used here to illuminate basic price policy issues provides a foundation from which to identify situations in which more complicated issues are likely to arise. An analysis based on the partial-equilibrium model for a commodity that uses a key input such as fertilizer in its production or faces an important crop substitute such as sugar will suggest that changes occurring in this commodity market would spill over into the fertilizer or sugar markets. If so, further analysis is required. This may mean analysis of one or more additional markets—fertilizer, sugar, or foreign exchange. It may require a model of the entire agricultural sector to capture the impact of more complicated timing, crop substitutions, labor market adjustments, and volumes of marketings within the agricultural sector and to the urban sector. It may ultimately involve at least a simple macroeconomic model to enforce consistency in national income accounts, trade flows, and savings and investment patterns.

There are a surprisingly large number of circumstances for which the partial-equilibrium model needs at least informal modeling and consistency checks to provide confidence in the robustness of results. Large price changes for an individual commodity, or significant price changes simultaneously for multiple commodities, are likely to cause multi-market spillover effects. Price changes for important commodities can be expected to have sectoral and macroeconomic effects. "Importance" is measured in one of four ways:

1. The commodity is the chief wage good in the society and forms a significant share—20 to 50 percent—of the average consumer's budget. This factor alone often makes food price policy analysis a maoroeconomic concern.

2. The commodity is a major source of farm income. This is important for two reasons: changing the price alters farm incomes and hence farm expenditures on goods and services that may have large employment content, and movements along the supply curve change input use and alter crop substitution possibilities.

3. The commodity is important to the country's international trade as an export or an import. Either way, changes in domestic prices are likely to alter trade volumes and hence the foreign exchange balance. In extreme cases, such as the very low prices paid by the government to cocoa farmers in Ghana, domestic price policy can nearly wipe out a country's foreign exchange earnings.

4. The commodity is important to the government budget, as either a generator of revenue or a significant drain because of large subsidies. The inherent need for governments to raise money through taxes has already been noted, and price policy analysis cannot ignore this need in favor of either efficiency or income distribution.

The most difficult examples of agricultural price policy analysis occur when the commodity is important for all four reasons—a common situation in developing countries in which a primary food staple is grown domestically and imported (or exported). Rice in Indonesia and Thailand, corn (maize) in Kenya, Zimbabwe, and Mexico, and wheat in Pakistan and Egypt, for example, easily meet these criteria. For these countries, the level of employment, the nutritional well-being of the poor, the extent of foreign exchange reserves, and the size of the budget deficit all hinge on the price of the critical commodity. Indeed, as violent political demonstrations over high food prices repeatedly make clear, even the existence of the government is intimately tied to price policy for basic foods. Analysis of these commodities extends the practitioner beyond the boundaries of partial-equilibrium models, even if the political dimensions remain outside the formal analysis itself.

The full extent of the economic adjustments to price changes depends not only on spillovers to other markets but also on the time frame of analysis, even when the analysis is restricted solely to the supply and demand curves for the single commodity. Producers and consumers tend to adjust more to price changes after several years than immediately after the change. The analysis here is for the "short run"—long enough for a first-round production response by farmers (one season or one year) and initial substitutions by consumers, but not long enough for full adjustment to take place. A useful rule of thumb is that about half the likely adjustment will take place in this initial decision period, with the remainder occurring in smaller increments in the following years. In these circumstances the "long run" is likely to be five to ten years.

Direct effects of the subsidy. In order for the government to establish its domestic price in Figure 2.4 at the desired level P_d, it must pay a per-unit subsidy equal to the difference between the world price and the domestic price ($P_w - P_d$) to the importers of the commodity. This subsidy can either be paid on a per-ton basis to commercial importers to cover their losses incurred by importing, or it can be covered by a budget transfer to a government food-importing agency to balance its books. Either way, government revenue from some source is needed to establish a domestic price of P_d to prevent the world price P_w from prevailing in the country's markets. This is what it means to have a price intervention in this direction: the use of general treasury revenue to keep domestic prices lower than the world price being paid to import the commodity.

The logic and analysis are entirely analogous for a country that keeps domestic prices lower than world prices, but which exports the commodity. In this situation, however, the import subsidy becomes an export tax. The government treasury earns revenues rather than disburses them, but the welfare transfers and efficiency losses associated with the example shown in Figure 2.4 are quite similar. There is, obviously, a major difference in the political economy of the policy, since the import subsidy drains the treasury, whereas the export tax adds to government revenues.

For a poor country in search of a tax base, such an export tax may be all that is available. In these circumstances, the efficiency losses are not real in the sense that the world price is not the true social opportunity cost to the economy if the government is to remain in existence. A direct implication is that any development assistance agency that encourages a country to "get prices right" must also help advise the government on how to increase direct tax collections. Note, however, that this consideration is irrelevant for price policies that require subsidies. The dynamic consequences of this difference depend on how the budget subsidies are financed or the export tax revenues spent.

The direct impact of the price subsidy can be evaluated in Figure 2.4. Consumers respond to lower prices by increasing consumption of the commodity; they move from point *d* to point *j* on the aggregate demand curve, and the total quantities they consume increase from Q_2 to Q_4. Farmers also react to the lower price, but in inverse fashion. They move from point *c* to point *g* on the aggregate supply curve as the lower prices force high-cost producers out of business; supplies forthcoming from domestic producers drop from Q_1 to Q_3.

Volume of trade. When the domestic price is equal to the world price, as in Figure 2.2, the country imports the difference between domestic production and total demand, or $Q_2 - Q_1$ in Figure 2.4. The subsidized domestic price causes this import gap to increase, as both increased demand and reduced supply contribute to larger import requirements. The new import gap is $Q_4 - Q_3$. How the original level of imports compares with the new level of imports depends critically on the elasticities of supply and demand. If they are near zero, quantities consumed and produced are little influenced by price, and the import gap does not change very much. If the response on the part of domestic decision makers is quite flexible, however, the changes in both production and consumption can be substantial. This is an empirical and not a methodological issue, and the empirical evidence shows that the relevant elasticities for individual commodities are non-zero even for a time horizon of one year and are very substantial on both the supply and demand sides in the long run.[1] Consequently, the import gap can be expected to widen significantly in response to increased price subsidies. Similarly, the volume of exports will drop significantly as a function of increased export taxation.

The extent of the change in imports is important because of

[1]Agricultural supply and demand elasticities are reviewed in several sources. The annotated bibliography of Timmer, Falcon, and Pearson is a useful starting point, as is its summary table of supply elasticities (p. 108). Some of the evidence on demand elasticities is reviewed by Scandizzo and Bruce (App. III). The distinction between supply and demand elasticities for individual commodities, for which substantial scope for substitution exists, and elasticities for aggregate food or agricultural output, for which response depends on more or less intensive use of inputs, is stressed in Streeten.

the expectations for budget planning of government officials. A natural tendency is to multiply *current* imports $(Q_2 - Q_1)$ by the desired price decrease $(P_w - P_d)$ to estimate budget revenues required to implement the policy; this is equivalent to area *cdih* in Figure 2.4. But examination of Figure 2.5 shows that the budget subsidy required would be much larger, $(Q_4 - Q_3)(P_w - P_d)$, or area *bejg*. Even if government officials are not concerned about the economic impact of the subsidy, they need to know supply and demand elasticities to prepare their financial projections. The time horizon of the analysis here is crucial. Producers and consumers may adjust very little in the first few months after the subsidy is implemented, and policy makers will note that their "naive expectations" were realized. But with time and increased experience on the part of producers and consumers with the new price environment, the adjustments shown in Figure 2.5 begin to take place, and these have obvious consequences for the government's budget.

Foreign exchange. Budget subsidies also alter the amount of foreign exchange needed to purchase imports. In Figure 2.5 the foreign exchange bill increases to $P_w(Q_4 - Q_3)$ from its initial value of $P_w(Q_2 - Q_1)$. Because of the government subsidy, the assumptions underlying the normative interpretation of the supply and demand curves no longer guarantee that the necessary resource adjustments to free up this added foreign exchange will be forthcoming. The price subsidy causes resource allocation to be inefficient, and some of the distortions induced may show up in foreign exchange shortages relative to the need indicated in Figure 2.5. For small changes in price and quantity this effect can probably be ignored, but otherwise it is likely to be an empirical issue requiring further analysis.

Market integration. Commodities freely traded in markets with equal access to buyers and sellers are largely fungible.

Figure 2.5. **"Expected" budget subsidies and foreign exhange costs relative to realized levels**

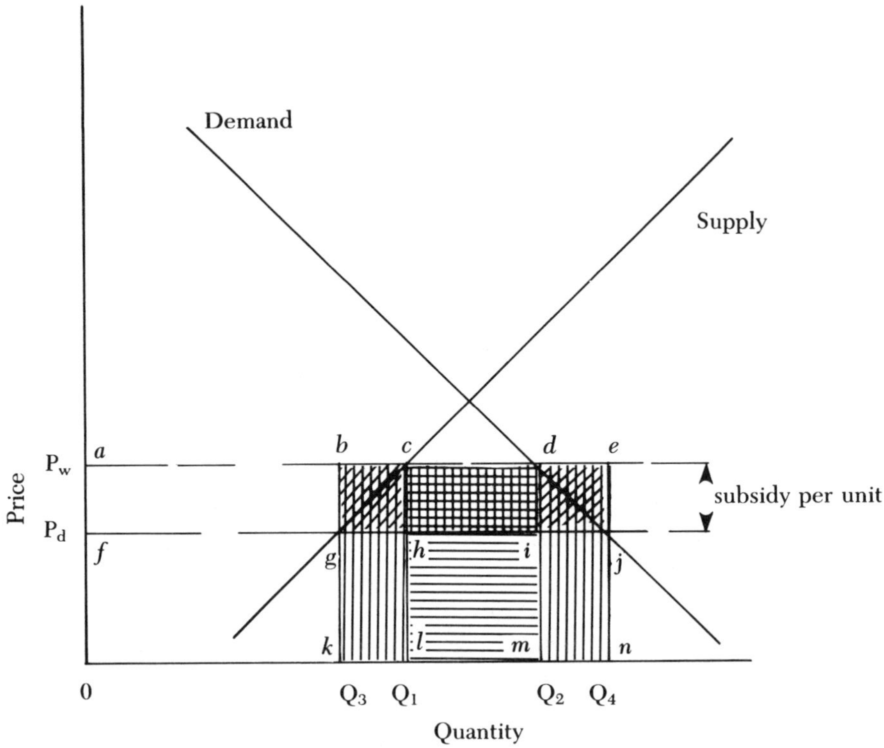

"Expected" foreign exchange requirements based on previous import levels (*cdml*)

Additional foreign exchange requirements because of adjustments to subsidized price by domestic producers and consumers (*denm* + *bclk*)

"Expected" budget subsidy needed to implement price policy, based on previous import levels and desired domestic price level (*cdih*)

Additional budget subsidy incurred because of adjustments to subsidized price by domestic producers and consumers (*deji* + *bchg*)

There is little difference between a ton of imported rice and a ton of domestic rice because the two commodities are very close substitutes in consumption. By subsidizing the price of the imported rice only, the market price of *all* rice can be driven down to the desired level P_d if quantities are unrestricted and trade among markets takes place, that is, if markets are integrated. This is a powerful result indeed and accounts for much of the popularity of price intervention as a tool of public policy in developing countries. By operating on relatively small amounts of easily controlled imports (or exports), the entire price structure for all the commodity produced and consumed within the country can be affected. This happens only because of arbitrage between markets; low-priced commodities tend to find their way to higher-priced markets, and thus prices in the first market are raised and those in the second market lowered. Where domestic commodity markets are isolated, unconnected, or fractionated by poor infrastructure, communications, or direct government interventions in marketing activities, the type of price policy analysis carried out so far provides a quite incomplete guide to the ultimate impact of price changes.

On the other hand, markets for food and agricultural commodities tend to be much more connected geographically and temporally than many government officials realize. A policy based on the presumption, for example, that a price for food can be implemented in the cities using imported grain with no impact on farmers in rural areas because of a "poor" marketing system (or because a government parastatal handles some of the marketing activities) would probably not work even in the short run. Almost certainly its implementation would break down in the long run as traders reacted to changed profitability of marketing grain.

Welfare effects. Unless governments purposefully isolate markets (and there is no guarantee they can), the subsidy policy is likely to have a reasonably uniform impact on prices faced by consumers and producers. This places a heavy, but often invisible, burden on producers. Consumers receive benefits quite explicitly and visibly through the budget subsidy needed to implement the lower domestic price, but they also gain at the expense of producers, who must pay an implicit tax of the same amount per unit. In countries where grain imports are only a small proportion of total consumption and production, the implicit tax on producers that is transferred to consumers would be larger than the visible budget subsidy. Where only 5 percent of domestic needs are imported and subsidized directly, for example, producers must "pay" consumers an amount roughly twenty times larger than the budget subsidy as a consequence of the lower price for that commodity. In Figures 2.4 and 2.5, the amount of this transfer from farmers that consumers actually receive is $Q_3(P_w - P_d)$, corresponding to area *abgf*. It is important to realize that consumers are receiving benefits from two sources—from the budget subsidy directly and from producers indirectly. Clearly, a major effect of price policy is to change income distribution between producers and consumers. For households that are both producers and consumers of the commodity, the story is more complicated. The effect is largely dependent on the share of the output marketed.

If price policy interventions affected only the distribution of incomes between producers and consumers, the policy debate would be largely political. Economists might be involved in analyzing the incidence of the taxes needed to finance the budget subsidy. Their attention would be drawn to the impact of additional imports on the foreign exchange rate, and hence on the price structure for the rest of the economy, but they would have relatively little to contribute to the debate over price policy. Indeed, the income transfer effects and the politics of budget allocations have dominated the formation of price policy in developed and developing countries alike, despite the presence of another major component of price policy impact that only economists can identify and quantify—the loss of economic welfare to both producers and consumers that is caused by distortions in the allocation of resources from their most efficient use in production and consumption. These losses can be identified only within the framework of efficient resource allocation that underlies the entire neoclassical market paradigm, one example of which has been used throughout this discussion in Figures 2.1 through 2.5.

Resource allocation. The potential for price policy to affect the efficiency with which an economy operates, and not just the division of the resulting output, is the primary reason for starting the analysis with the simple partial-equilibrium model. This model shows dramatically both the income transfers and the efficiency losses that arise when resources are diverted from their best use. It is incomplete, however, and market spillover effects, dynamic effects, and the less-than-perfect market setting in most developing countries forbid firm policy conclusions made solely on the basis of the model. The skilled analyst can tease these issues out of the basic model, however, because it provides clear signals on where to look. The policy debate, then, has a common starting point for all participants; the arguments can be about the important issues, not the mechanics of the model.

The subsidy on imported rice used so far as an example results in efficiency losses, which are shown in Figure 2.6 as shaded triangles. Producers provide an implicit income transfer to consumers of an amount equal to the area *abgf*, or domestic production times the per-unit subsidy. Actual farmers' losses resulting from the drop in price from P_w to P_d, however, are greater than this amount by the area of the triangle *bcg*. The amount represented by this triangle, plus the area *abgf*, mea-

Figure 2.6. Efficiency losses in production and consumption with a subsidy on imports

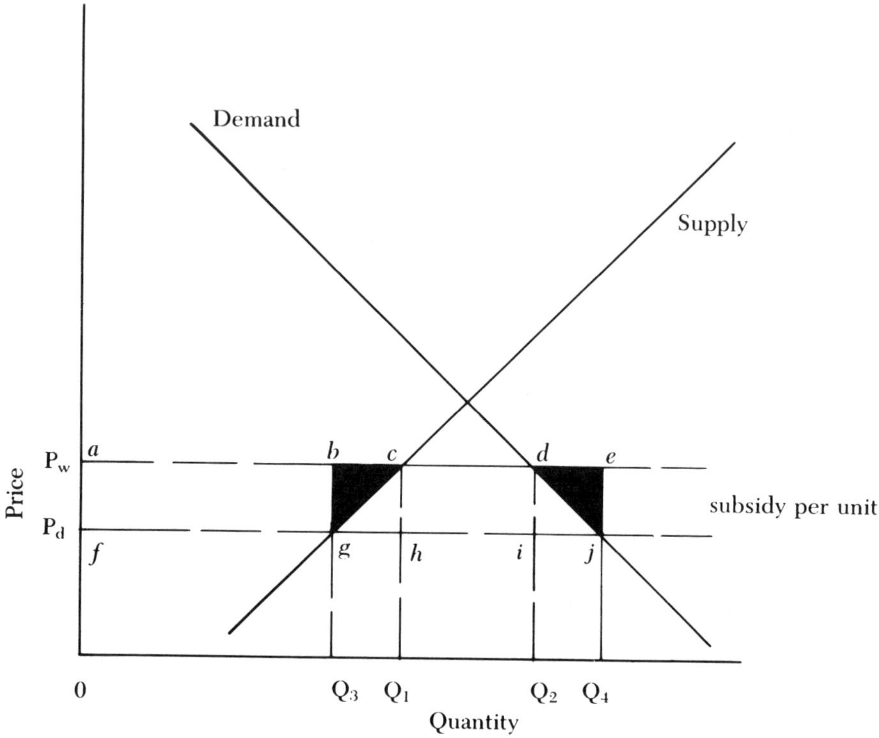

sures the total loss in producer surplus, or profits, caused by the new price P_d. At the world price P_w, domestic resources could have been used at costs lower than P_w to produce the commodity, and the output would have sold for P_w. The difference between the resource cost, which is measured by the supply curve itself (under the plausible assumption that individual farmers would not continue to produce a commodity to sell at less than its marginal cost of production), and the selling price is the producer surplus. This surplus or farmers' profit is measured in Figure 2.6 by the area above the supply curve and below the market price.

Part of the total loss in producer surplus (area *bcg*) is *not* accompanied by a commensurate gain in consumer welfare—otherwise the discussion would return to the "simple" issue of income distribution. Consumers have already received from the budget subsidy a gain in welfare corresponding to *bcg*, and so the additional loss to producers is a "dead-weight efficiency loss"; that is, somebody's welfare is reduced because of the misallocation of resources (expensive imported grain is being substituted for domestic resources that could produce it more cheaply), but no one else benefits from this particular loss.

Economists view such efficiency losses, especially on the production side, with genuine concern. At least in the simple world of Figure 2.6, all of the consequences of price policy for income distribution that are desired on political grounds can be achieved with subsidiary tax and direct income transfer programs, without incurring the efficiency losses. Unfortunately, such taxes and direct transfers are difficult to implement for political and administrative reasons; unlike the implicit income transfers effected by price policies, they must be made explicit and channeled to individuals through a government bureaucracy. Such direct fiscal approaches create their own inefficiencies and disincentives to expand output, and competing interest groups become involved in the political debate over appropriate price policy.

As Stiglitz has demonstrated, the efficiency of market economies holds only for circumstances that include a complete set of markets and perfect information. With the imperfect information and incomplete set of markets that characterize developing countries, a combination of taxes and subsidies can actually improve overall efficiency and make everyone better off (i.e., achieve a Pareto improvement).

> This implies that there is a *potential* role for the government. That is, the government could effect a Pareto improvement if (i)

> it had sufficient knowledge of the structure of the economy; (ii) those responsible for implementing government policy had at least as much information as those in the private sector; (iii) those responsible for designing and implementing government policy had the incentives to direct policies to effect Pareto improvements, rather than, for instance, to redistribute income (either from the poor to the rich or vice versa, or from everyone else, to themselves), often at considerable loss to national output. Informational problems, including incentive problems, are no less important in the public sector than in the private; the fact that we have studied them well in the latter does not mean that they are not present in the former. The consequence of these remarks is to make us cautious in recommending particular government actions as remedies for certain observed deficiencies in the market. (Stiglitz, pp. 9–10)

Economists have a further concern over the impact of allocative efficiency losses. Those in the economics profession have understood for many years that the triangles of allocative losses are small relative to the rectangles of income transfers. Similarly, the triangles are likely to be small relative to technical or X-efficiency losses caused by a failure to achieve the maximum output from a given set of inputs. Consequently, static allocative inefficiencies have not been the cause of much concern until recently because they seemed likely to be small relative to both the direct effects on income distribution and the losses in technical efficiency. Development economists have concentrated on transferring modern ("efficient") technology to developing countries and on creating an income distribution that could generate a large volume of savings accessible to formal financial intermediaries.

These have turned out to be the wrong concerns for fostering a sustainable development process. Resource allocation is an inherently dynamic process; and the static efficiency of allocation is an important element in creating dynamic efficiency, that is, in the rate of economic growth. Research at the World Bank has documented, in its *World Development Report 1983*, a statistical relationship between distortions in prices and rate of growth in gross domestic product between 1970 and 1980 for a sample of thirty-one developing countries. Three categories of prices were used in the analysis: foreign exchange pricing, in which exchange rates, protection of manufacturing, and agricultural taxes or tariffs were included; factor pricing, in which labor and capital prices were included; and product pricing, in which power tariffs and inflation rates were included.

There are many dimensions to price distortions, and the "composite distortion index" used in the World Bank's analysis captures them only crudely. Likewise, there are many reasons for the varying rates of economic growth, including the luck of natural resource endowments or prices in world markets. With all the provisos, however, the evidence is clear that price distortions not only cause the sort of short-run, static inefficiences shown in Figure 2.6 but also contribute powerfully to the varying performance in economic growth observed in the 1970s. As a consequence, economists are returning to their concern over price policy, even in a political economy context in which the ideal tax and direct transfer programs are ruled out. Chapters 6 and 7 discuss this macroeconomic and dynamic impact of price policies as well as their consequences for economic growth.

Figure 2.6 also shows a second source of allocative losses—the consumer efficiency loss. Budget subsidies on behalf of consumers correspond to area *bejg*, but consumer welfare increases only by area *bdjg* (in addition to the transfer from producers of area *abgf*). The difference is the area of triangle *dej*, which is an actual budget expense but for which consumers receive no benefit. The reasons are analogous to the losses in producer surplus corresponding to triangle *bcg*, but the consumer welfare losses are not quite so real and tangible as the measurable monetary losses incurred by farmers when prices drop.

Consumers are made better off by lower prices for a commodity, and the traditional way of illustrating, even measuring, that improvement in welfare is through the change in "consumer surplus." This change is shown in Figure 2.6 by area *adjf*—the area under the demand curve bounded by the price before and after the price policy intervention.

This gain in consumer welfare is provided by measurable transfers of economic resources to consumers: *abgf* from losses to farmers and *bdjg* in budget subsidies. Just as farmers' losses are larger than consumer gains by the amount of the efficiency loss *bcg*, so too are total budget subsidy costs plus producer transfers larger than the gain by consumers, in this case by the triangle *dej*. This area represents the consumer efficiency loss. In analogous fashion to the producer efficiency loss, the magnitude of the consumer efficiency loss is a direct function of both the size of the price "distortion" (the difference between the world price P_w and the domestic price P_d) and the elasticity of the demand curve with respect to price. If consumers do not change their consumption patterns when prices change, no efficiency losses are incurred. Only the income and budget transfers remain as price policy effects. As consumer response gradually increases (especially over time as consumers gradually make substitution decisions more consistent with the new price environment), the allocative efficiency losses are magnified.

Consumer efficiency losses are also likely to have dynamic allocative effects on an economy. The evolving patterns of demand over time provide the main driving force for investment and structural change. Subsidizing important commodities, especially an imported basic foodstuff (particularly if provided by food aid), can drastically skew development patterns and retard growth in employment and aggregate economic output. At the same time, however, the additional consumption of the basic foodstuff made possible by the subsidized price might have important consequences for nutrition and development of human capital, effects that complicate the welfare analysis of price policy. These consequences are dealt with directly when the analysis is disaggregated by income class in Chapter 5.

The price of a subsidy. Within the static, partial-equilibrium framework of neoclassical economics, what is the overall judgment with respect to the impact of a price policy intervention that lowers the price of an imported foodstuff from the world market price P_w to a policy-determined price P_d? Tsakok works out the algebraic steps needed to quantify the answer, but Figures 2.4 through 2.6 provide an adequate intuitive understanding of the issue. Producers lose by the amount of area *acgf*, of which *abgf* is transferred to consumers and *bcg* is sheer dead-weight efficiency loss, benefiting no one. Consumers gain by the amount *adjf*. The budget must pay a subsidy from general government revenues equal to *bejg*. The budget subsidy plus the income transfer from producers to consumers also exceeds the gain in consumer surplus by the amount *dej*—a dead-weight efficiency loss on the consumer side. Imports of the commodity increase from $Q_2 - Q_1$ to $Q_4 - Q_3$ because production falls and consumption rises, and so foreign exchange requirements rise in direct proportion. These imports may be a relatively small fraction of total production and consumption, in which case the total budget subsidy needed to drive down the price from P_w to P_d would be quite small in comparison with the implicit transfer of income from producers to consumers. This transfer shows up nowhere in formal economic accounts.

The increased volume of imports may contribute to decreased food security for the country in question if foreign exchange is not readily available to the importing agencies in the case of sudden price increases on world markets. But food imports do not necessarily imperil a country's food security. A well-managed food import component to an overall food policy that actively uses international trade as a source of growth is likely to improve food security. In many countries the variance of local production is much greater than the variance in the level of supplies available from world markets (but not necessarily the variance in their prices). Since providing a smooth flow of food for consumption from domestically supplied buffer stocks is normally more expensive than using trade to guarantee supplies, food security may well be reduced as a country imposes self-sufficiency in its basic food grain.

The partial-equilibrium model has identified the "price of admission" to analysis of the subsidy, but not the full "price" of the subsidy itself. The story continues in later installments: in Chapter 4, where feedback effects on foreign exchange rates and world market prices are considered; in Chapter 5, where the analysis is disaggregated to look inside households with different income levels; in Chapter 6, where the initial market effects spill over into connected markets for inputs, substitute outputs, factors, and ultimately to macroeconomic consequences; and in Chapter 7, where dynamic and political economy factors are examined. Adding all of these additional factors into the analysis does not necessarily add to the "price" of the subsidy because these factors influence government objectives in a variety of ways. The net result is not discernible from this or any other model. It will depend on the unique circumstances of the setting and on the objectives themselves.

Trade Policy to Protect Farmers

Not all countries discriminate against their rural sectors by subsidizing food to consumers at the expense of farmers. Many governments, especially in developed countries, use policy interventions that impose trade barriers against imports (or subsidies to exports) to raise prices and provide added incentives for domestic producers, especially when world prices are very low. The same framework used to analyze subsidies can be used to

Figure 2.7. Price formation with an import tariff or quota

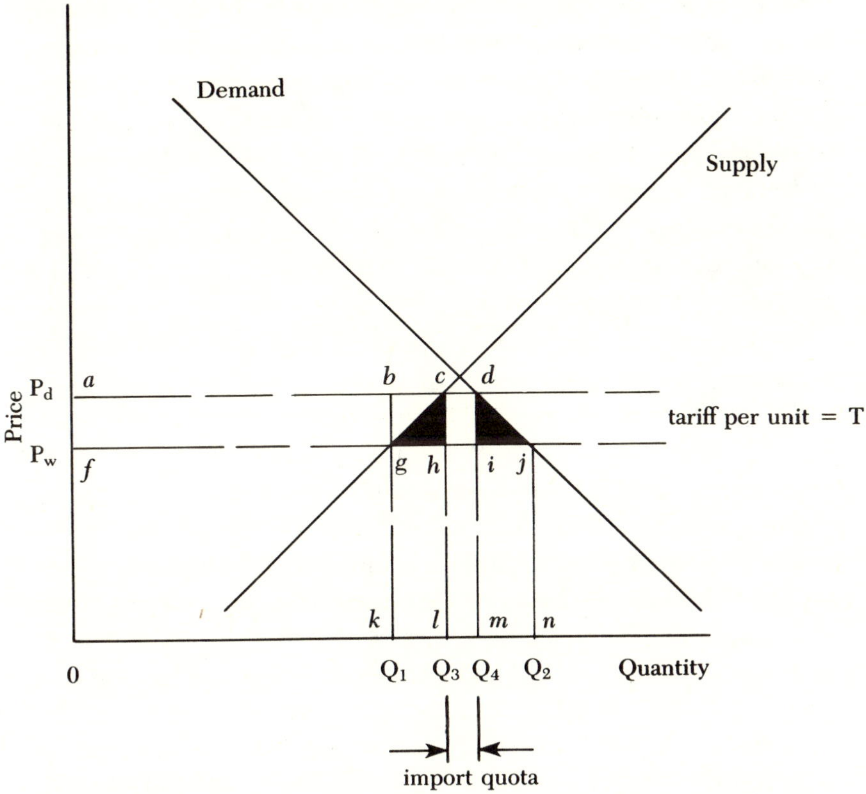

analyze the effects of these policies. Figure 2.7 uses the same basic assumptions and starting point with respect to the world price to examine such interventions.

As in the analysis of a food price subsidy, the initial conditions begin with the domestic price of a commodity equal to the world price for both producers and consumers (with appropriate allowance made for marketing costs). Domestic production is equal to Q_1, consumption is equal to Q_2, and supplies of $Q_2 - Q_1$ are imported from the world market at price P_w to make up the difference. In this example, government policy responds to a different set of interest groups or recognizes that food costs are no longer such a critical item in the consumer budget. Rural concerns over low prices and foreign competition then outweigh consumer concerns over high prices, and the aim of policy is to increase the commodity's price (see Anderson and Hayami). The policy intervention can take one of two forms: either a per-unit tariff can be placed on the imported commodity ($P_d - P_w = T$ in Figure 2.7), or import quotas can be imposed to drive the market price up to the level desired.

In Figure 2.7 the desired price P_d can be achieved by reducing imports from $Q_2 - Q_1$ to $Q_4 - Q_3$. Total supplies in the domestic market then just equal demand at the new price P_d. It is important to note that both producers and consumers react to the higher price by adjusting their behavior in appropriate directions. This adjustment makes the actual implementation of price P_d more difficult if quantitative controls are to be used (via import decisions by a parastatal, for example) than if a simple per-unit tariff is imposed. From the perspective of maintaining a stable domestic price at P_d, the tariff has the disadvantage that any fluctuations in the border price P_w will be directly transmitted to the domestic price P_d, in contrast to quantitative import quotas of $Q_4 - Q_3$, which will maintain P_d no matter what happens to P_w. As noted above, full information about producer and consumer elasticities is needed to use quotas to fine-tune domestic price policy if existing private marketing agents are used to implement the quantitative controls. Few developing countries have access to reliable elasticities—which are not very stable anyway—and most use government agencies or parastatals to implement their price policies directly.

With the new, higher price P_d, producers gain because they receive higher returns to current production, $Q_1(P_d - P_w)$ or area *abgf*, and also because they now find it profitable to expand output from *g* to *c* on the supply curve (from Q_1 to Q_3 in aggregate output). The additional profit, or producer surplus, accruing to the farmers thus corresponds to *bcg*. Total gain by farmers is *acgf*.

Consumers lose when the price goes up, and their loss is measured by the change in consumer surplus under the demand curve. This total loss is shown as area *adjf*, an area considerably larger than the gain in farmer welfare. Part of the difference accrues to the government budget in the form of tariff revenue. The size of the budget surplus—the mirror image of the budget subsidy needed in Figure 2.5 to lower prices relative to the world market price—is determined by the per-unit tariff times the volume of imports, $(P_d - P_w)(Q_4 - Q_3)$, which is equal to area *cdih* in Figure 2.7. Again, it is important to note that budget revenues will be sadly under the estimates if planners assume that revenues equal *current* imports $(Q_2 - Q_1)$ times the newly imposed tariff. Depending on the elasticities of supply and demand and the initial volume of imports, the actual revenue receipts from tariff collections could be quite small relative to these initial estimates.

Such adjustments by producers and consumers occur whenever a tax is imposed on an economic activity. Even lump-sum profit or income taxes affect decisions with regard to labor and leisure or savings and investment. Since some taxation is essential, however, one of the goals of public finance is to minimize such distortions, not to eliminate them altogether.

In some cases the distortions can be severe enough to change budget revenues into budget subsidies. If the desired price P_d should turn out to be *above* the equilibrium point at which domestic production equals domestic consumption, a subsidy would be needed in order for the surplus production to be exported to the world market at price P_w. The European Economic Community and Japan understand this problem extremely well. It happens frequently because the longer-run elasticities of supply and demand are assumed to be much smaller than they turn out to be; assured high prices induce substantial research and farm-level investments that raise productivity, and consumers gradually find alternative substitutes.

Even with the budget surplus counted in Figure 2.7, the total loss in consumer surplus still exceeds the total producer gain plus budget revenues. The figure shows that two triangles remain to be accounted for: area *chg* under the supply curve, and area *dji* under the demand curve. These are dead-weight efficiency losses in production and consumption, respectively. The production loss *chg* occurs because resources equal in value to this triangular area are used to produce the commodity over and above the resources that would be needed to import it. Imports cost only the area *ghlk*, whereas domestic resources equal to this area *plus* the triangle *chg* would be used to produce the commodity. It is double counting to argue that the imports require foreign exchange, whereas the domestically produced output does not. In a smoothly functioning economy the domestic resources represented by *ghlk* will be able to *generate* the needed foreign exchange.

On the consumer side, the loss in consumer surplus shown as area *dji* is not matched by an equivalent income gain to either producers or the budget. The loss in consumption from Q_2 to Q_4 is matched by savings in resources used to purchase the imports equal to area *ijnm*, but the consumer surplus area *dji* is a dead-weight efficiency loss in consumption.

The Partial-Equilibrium Perspective on Price Policy Interventions

It is easy to see from the preceding examples why economists are so concerned to "get prices right." Price distortions, as measured by deviations of domestic market prices from appropriately defined border prices, cause inevitable losses in economic welfare because of dead-weight efficiency losses. Within the fairly narrow assumptions of the neoclassical model, these losses arise whether prices are raised or lowered and whether the commodity is imported, exported, or switched from one to the other because of producer and consumer response to price changes. Efficiency losses can be avoided only when response elasticities are zero. In such a world, prices do not matter at all for resource allocation. They are important only as a key element in income distribution.

When the commodity in question is the basic food grain for the society, the income distribution effects of price policies can be extraordinarily powerful. Indeed, the government may have no other policy instrument capable of putting real purchasing power in the hands of consumers (especially urban consumers) that has nearly the impact of a price subsidy on imports. It is perhaps no wonder that many government officials choose to believe that supply and demand elasticities for such a commodity are nearly zero; if this were true, there would be only political (and positive) consequences to food price subsidies, with no economic distortions and no serious reductions in growth opportunities.

Figure 2.8 summarizes the effects of policy interventions that change the market price of a commodity. Dual price policies that subsidize only producers or only consumers are not included in the figure. Market prices are presumed to affect both. Although the analysis so far has not examined the impact of a price policy intervention on an exported commodity such as coffee or cocoa, the results are entirely straightforward. The implications of price interventions for both imports and exports are shown in the figure.

Although Figure 2.8 divides the world neatly into four

Figure 2.8. **Summary of price policy effects**

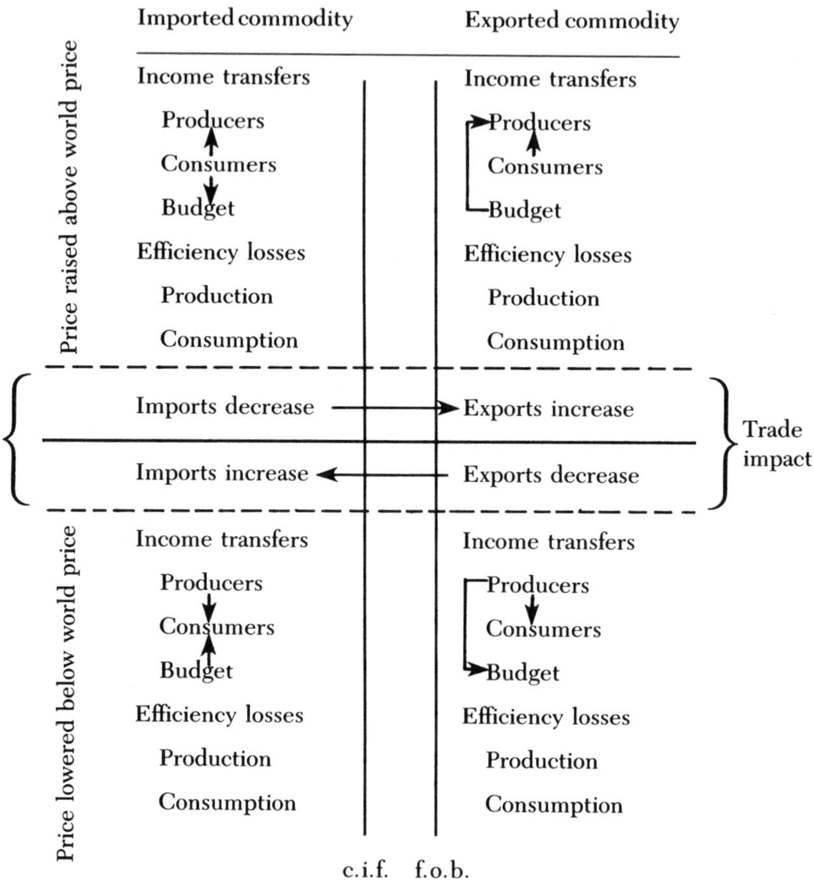

categories—domestic prices are either higher or lower than world prices and the commodity affected is either imported or exported—the real world is not quite so accommodating. Policies themselves can determine whether a country imports or exports a particular commodity. If supply and demand elasticities are quite substantial (perhaps because of induced innovation and investments) and the difference between domestic and world prices is maintained consistently for several years, imports might decline to zero, for example, and exports would be needed to dispose of the surplus. Similarly, sufficient taxation of an export commodity can eventually lead to imports being required to maintain the policy-determined price levels. These potential "switches" are depicted in the figure by arrows linking the trade effects of price policies.

Figure 2.8 illustrates the important dynamics between price policy and budget revenues. Imported commodities whose domestic prices are raised above world prices (upper left) and exported commodities whose domestic prices are pushed below world prices (lower right) earn budget revenues for the government. If the price changes are substantial, that is, if the tax or tariff rates are high, the adjustments by domestic producers and consumers can be large enough to shift the commodity from an import to an export, or from an export to an import. Important examples of the former case are those of rice in Japan or Indonesia. Few countries can afford to convert an export-earning and revenue-generating commodity into a subsidy-paying imported commodity, although corn and wheat in Mexico and corn in several African countries have come close to being such examples.

Such switches from exporting to importing, or vice versa, can have a dramatic effect on a domestic agricultural system, especially on marketing activities because the direction of trade is reversed. The impact of price policy on the marketing sector and the problems of actually implementing a price policy are the topics treated in the next chapter. In addition, because world commodity prices are very unstable, the domestic price could remain unchanged from one year to the next, but the policy *impact* as measured by standard price analysis would shift from the upper to the lower part of Figure 2.8, or vice versa. Even with stable domestic prices the apparent impact on producer or consumer welfare could be inverted if world prices rise above or fall below the domestic price. This obviously raises the question of what border price to use for price policy analysis, which is the topic of Chapter 4.

3

Implementing Price Policy: The Impact on Markets and Marketing

Price policy is implemented by changing the quantity or distribution of a commodity in markets. Sometimes only the engineering functions of marketing—storage, transportation, and processing—are carried out in markets, and price formation is reserved entirely for government policy and government agents. All economies must also facilitate the exchange of ownership of commodities from farmers to consumers, a function performed by the marketing sector and marketing agents, whether private or public. In market economies, markets not only perform the engineering functions but also serve as the locus of price formation, which may also be influenced by direct or indirect intervention by the government.

Price Policy and Markets

Price formation has three important aspects: (1) the determination of the domestic price to sellers and buyers of a commodity relative to its border price and relative to substitutes and complements; (2) the stability of that domestic price over time, given that border prices for most agricultural commodities fluctuate considerably; and (3) the price margins formed for a commodity between points in time, which affect storage decisions, between points in space, which involve transportation decisions, and between forms, which reflect degrees of processing.

It is possible for price policy to affect only the first of these topics—the domestic price level. A fixed tariff per ton, for example, would keep the domestic price of an imported commodity higher than the border price by the amount of the tariff but would permit domestic price variations equal to those in international markets. Similarly, the structure of price margins across space, time, and form would not necessarily be influenced by the wedge between domestic and border prices. Such a "pure" price policy is usually implemented for revenue purposes rather than for its effect on the domestic agricultural system. Price policies designed to influence participants in that system, whether consumers, producers, or both, almost inevitably intervene in price stability as well as price level and thus significantly affect the formation of price margins.

In market economies these prices influence the allocation of resources in both production and consumption. The coordinating role of prices determined in competitive markets generates the efficiency claimed for market economies. These claims have substantial merit, but actual market performance and its welfare impact depend critically on just how efficiently marketing activities are carried out and thus how efficiently markets generate and transmit price signals.

All of these issues are connected. Price policy, interpreted as interventions to alter domestic prices relative to border prices and to increase the stability of domestic prices, influences price margins and the allocation of resources. The price margins in turn influence private and public decisions with respect to the food marketing activities. How efficiently these activities are carried out is determined by comparing the costs of storage, transportation, and processing with the price margins actually generated in the various commodity markets. How efficient these markets are in matching costs with price margins during competitive price formation then determines whether the allocation of resources in a market economy generates an efficient outcome. It is no exaggeration to say that the links between price policy and food marketing activities take the food policy analyst to the very core of an economy and the most basic issues concerning the consequences of market organization for economic efficiency and income distribution.

Impact of Price Policy on Marketing Activities

Most food and agricultural price policies are designed to influence price margins as well as price levels, although the subsequent effect on marketing participants is usually not carefully considered. Each of the three basic marketing activities—storage, transportation, and processing—is influenced by price policy. Storage decisions are usually most affected in the short run, and this chapter looks especially at the influence of price policy on decisions to store commodities and the role of public agencies relative to private traders in this activity.

Storage

The introduction of a ceiling price as the specific manifestation of a price policy obviously affects expectations about future prices on the part of a private trader engaged in purchasing commodities at harvest for sale at a later season when prices are higher. If the ceiling is effectively enforced only in urban markets through supplies of imported grain distributed (with a subsidy if necessary) by a national logistics agency, the private sector will leave much of the task of provisioning urban markets

Figure 3.1. The structure of urban and rural rice prices with an urban ceiling price policy

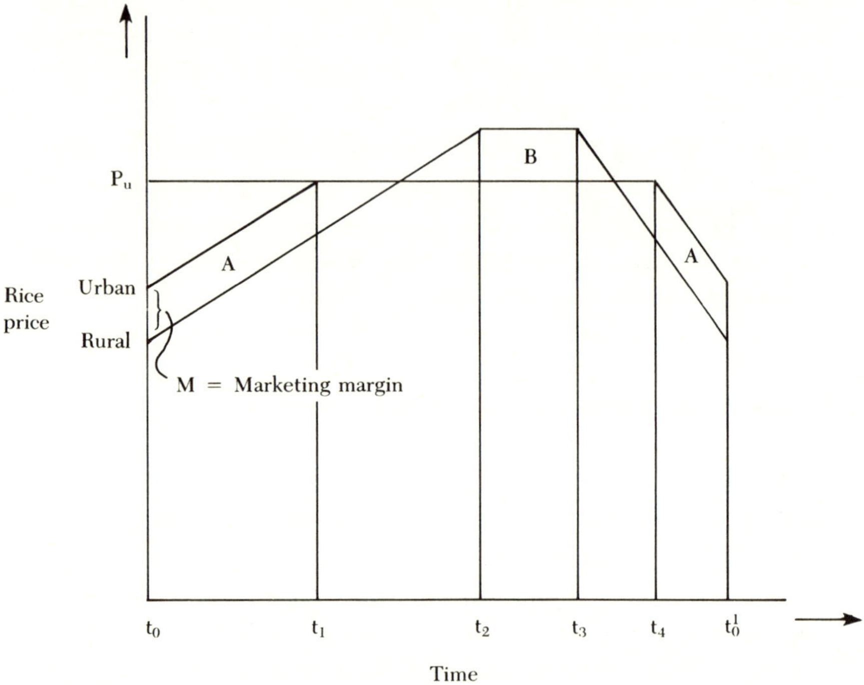

Source: C. Peter Timmer, "A Model of Rice Marketing Margins in Indonesia," *Food Research Institute Studies*, 13, no. 2 (1974), p. 151.

to the public sector (a task examined below). But what happens to grain marketing in rural areas?

This question is addressed by Figure 3.1, which shows a schematic framework which has been used successfully in several settings (Timmer, 1974; Pinthong). The horizontal axis shows time from one harvest period to the next, and the vertical axis measures rice prices. Urban price formation is determined by rural prices plus a constant marketing margin M, until the urban ceiling price P_u becomes effective, at which time the public marketing agency must assume primary responsibility for providing rice supplies to urban markets.

In rural markets prices are free to rise to cover the full costs of storage, even when they exceed the urban ceiling price. If they rise far enough, by the marketing margin M, for example, it becomes profitable for private traders to ship rice from urban markets *back* to the rural areas. In this setting, the urban ceiling price sets a rural ceiling price higher by the margin M rather than lower by the margin M. Such a pattern of price formation requires a break in physical links between rural and urban markets; this break occurs from t_1 to t_2 and from t_3 to t_4.

Without the price policy, urban prices would continue to rise until the new harvest begins at t_3, and a constant margin would separate rural and urban prices all year long. The price policy has a dramatic impact on the structure of rice marketing: it changes the participants in urban marketing from entirely private to significantly public, and it introduces reverse flows of rice from the traditional rural-to-urban pathways.

The relative balance between private- and public-sector marketing agents can be seen even more clearly in a model that focuses specifically on the urban food market but in which both a floor price P_f and a ceiling price P_c are defended by the public logistics agency. The model, illustrated in Figure 3.2, was originally designed to understand the impact of alternative margins between P_f and P_c in an Indonesian context. The top panel presents the "base case," with no price policy interventions. Seasonal price formation thus reflects the full costs of storage, starting at P_l when harvest at time t_0 is completed and rising to P_h when harvest at time t_1 begins. The horizontal axis is scaled to reflect the *size* of the harvest sold for storage, H, so areas represent total revenues. The rectangle HP_l is gross revenues to farmers, and the shaded triangle represents total storage costs paid by consumers. If unit storage cost S is equal to $P_h - P_l$, this total storage bill for consumers is $T_s = \frac{1}{2}HS$.

The bottom panel of Figure 3.2 shows the results of introducing a floor and ceiling price policy which benefits both producers and consumers: the floor price is higher than the previ-

Figure 3.2. **Impact of price policy on the role of the private marketing sector and the size of government subsidies**

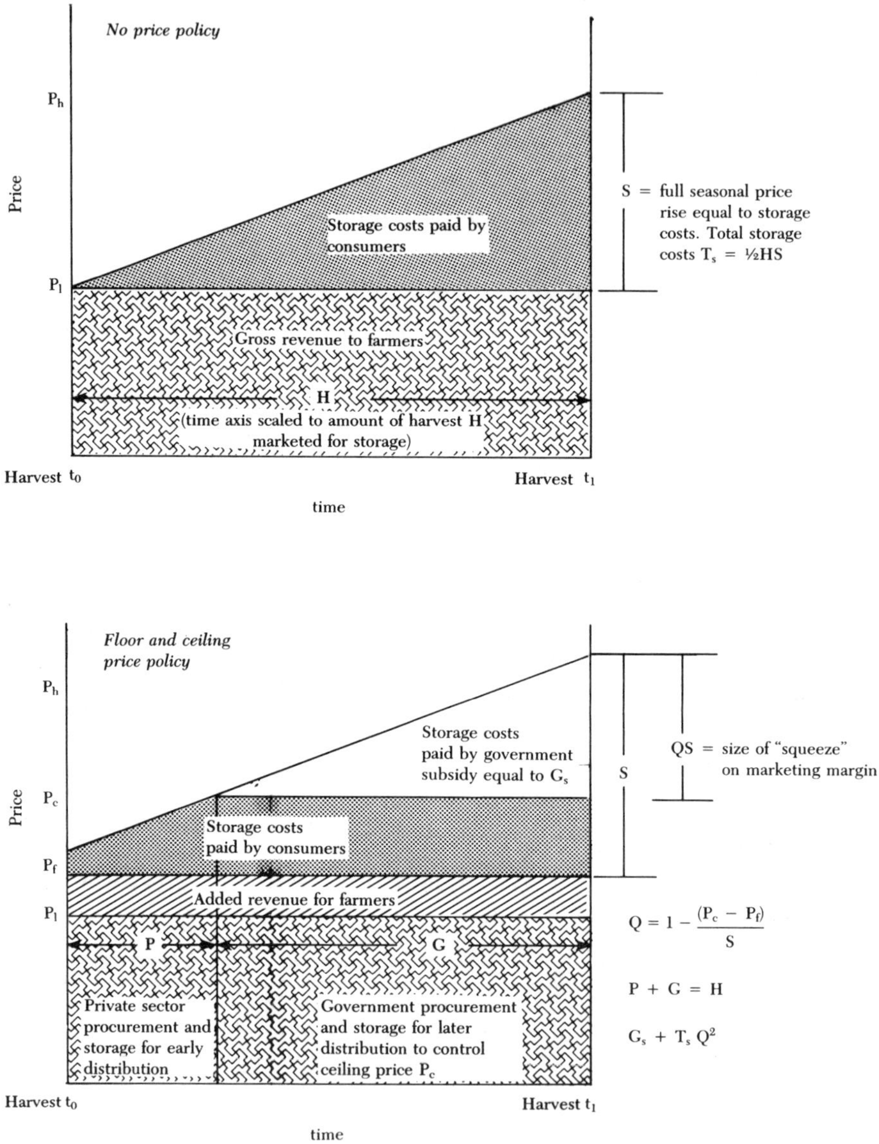

ous harvest price ($P_f > P_l$), and the ceiling price is lower than the previous preharvest price ($P_c < P_h$). Although the figure shows farmers and consumers both receiving a price benefit, it is clear that farmers have much larger total gains since their price gain applies to their total sales, whereas consumers' gains apply to only a small fraction of their purchases. Consumers' total costs are in fact higher because of the higher starting point for prices. Different price assumptions could, of course, alter this distribution of benefits.

Two important consequences flow from the implementation of the price policy shown in the bottom of Figure 3.2. First, since the unit storage cost *S* represents real economic resources, a government subsidy must be paid to the food logistics agency responsible for carrying out the floor and ceiling price policy. This might be an explicit line item in the government budget, or it might come through subsidized interest rates, access to government-owned warehouses at low rates, and so on. But the figure shows that this subsidy can be a quite substantial share of total storage costs. The government subsidy incurred is $G_s = T_sQ^2$, where Q is the proportion of total storage costs *not* covered by the seasonal rise in prices. One can think of Q as the proportional "squeeze" on the profit margin of any private traders attempting to store rice for later sale, and QS is the "squeeze" per unit of commodity.

It is perhaps surprising that the total government subsidy required to implement the floor and ceiling price policy is a function of the square of the proportional squeeze rather than a linear function, but the result is intuitively clear from the figure.[1] This result is extremely important, for it means that policy efforts to benefit both farmers and consumers by squeezing the

[1]The formal geometric proof is also straightforward: $G_2 = 1/2\ G(QS)$; since $G/H = (QS)/S$, by the rule of similar triangles, $G = QH$. Substituting for G, $G_s = 1/2\ (QH)(QS) = 1/2\ HSQ^2$. Since total storage costs $T_s = 1/2\ HS$, $G_s = T_sQ^2$.

marketing margin will require subsidies which rise with the square of the squeeze on the margins. Such unexpectedly rapid increases in subsidy costs often catch logistics agencies and finance ministries unprepared and result in failure to implement fully the announced floor and ceiling price policies. These failures immediately affect the expectations of all market participants—farmers, traders, and consumers—and can rapidly lead to speculative activity which further limits the government's efforts to set prices.

This impact on private traders conveys the second important message in Figure 3.2. With no price policy, the private sector carries the full burden of buying and storing grain, provided its full costs of storage, including returns to capital, risk, and managerial skills, are covered. Competition among private traders is required to prevent accrual of monopoly profits and exploitation of farmers and consumers.

Does the government provide such competition when it introduces a floor and ceiling price policy? It does if the seasonal price rise $P_c - P_f$ reflects the full *competitive* costs of storage. Figure 3.2 shows what happens if the permitted seasonal price rise is *less* than competitive costs. The government agency itself must then carry out some of the marketing activities directly, and its task is directly proportional to the price squeeze Q. Within the simple analytics of Figure 3.2, the government agency must thus procure and store for distribution a share of the marketed harvest equal to Q, the price squeeze. When $Q = 0$, the private sector does all the marketing; when $Q = 1$, the government must do it all. There can be no clearer example of the dramatic impact of price policy on food marketing.

Transportation

Price policy has spatial dimensions which are analogous to the temporal ones. Floor or ceiling prices that are implemented uniformly over a wide geographic region are sure to have direct effects on the role of the private sector in transporting commodities from areas of low prices to areas of high prices and in earning a profit from the spatial arbitrage (in addition to affecting the structure of production and consumption). With uniform floor prices the relatively favored areas close to points of final demand will be serviced by the private sector, and more distant areas will be left to be handled at high cost by the government agency. Since such agencies seldom have real operating costs lower than those of the private sector, these long-distance activities require subsidies.

An analogous point holds when a uniform price floor is announced for all qualities of a commodity but retail prices reflect quality differentials. Again the government agency will end up defending the floor price by buying the lowest quality offered of the commodity. Public agencies usually incur storage costs higher than those of the private trade because low quality in food grains is related to high moisture content and insect damage. These cause substantial losses in storage.

Price policy can also directly affect the regions of a country that are competitive with imports. For a given cost of production, the landed import cost (after tariff or subsidy) minus transportation costs from the producing hinterland traces out a "competitive contour" beyond which domestic production incurs higher total costs in supplying consumption needs in the port than do imports. Such contours determine the nature and direction of marketing activities. As prices vary because of policy changes, the location of the contour shifts as well. Traders must then adapt their marketing activities; they can go farther inland in search of supplies if the import price rises, and they are restricted to nearer the port if prices fall.

As noted in Chapter 2, some price policies are sufficiently powerful that they can cause a switch in direction as well as distance for food marketings. If import prices are low enough or subsidies large enough, it becomes profitable to market grain from the ports to the producing hinterland, and immense competitive pressures are thus put on domestic producers. A budget crisis (and a consequent reduction in the subsidy on imports) or a currency devaluation can sharply alter these prices and thus raise incentives to farmers and provide profitable opportunities to market domestic grain in the port cities. Switching the whole direction of flow for a country's marketing system, however, is no easy task and will be only partially the result of price policy interventions. Both a public agency and private traders need considerable time to adjust before efficient marketing can be expected.

Processing

Price policy probably has relatively smaller impact on processing activities than on storage or transportation, but even here the effects can be significant. Figure 3.3 shows a flow diagram of factors influencing the choice of technique in rice milling, a choice that has dramatic consequences for rural employment, the efficiency of conversion between paddy and milled rice, and the quality of rice available to consumers (Timmer, 1972). The figure charts the conversion of unhusked rice (paddy) to milled rice by one of five different technical milling processes: hand pounding, small rice mills, large rice mills, small bulk facilities, and large bulk facilities. Each process has its own investment cost and number of workers needed per shift, which determine the capital and labor costs of milling rice. The number of workers required determines the level of employment in the rice milling sector; in the Indonesian example each process was roughly twice as capital-intensive and created half as many jobs as the one ranked before it in capital intensity. Large rice mills thus required, for example, about

Figure 3.3. Factors influencing the choice of technique in rice milling

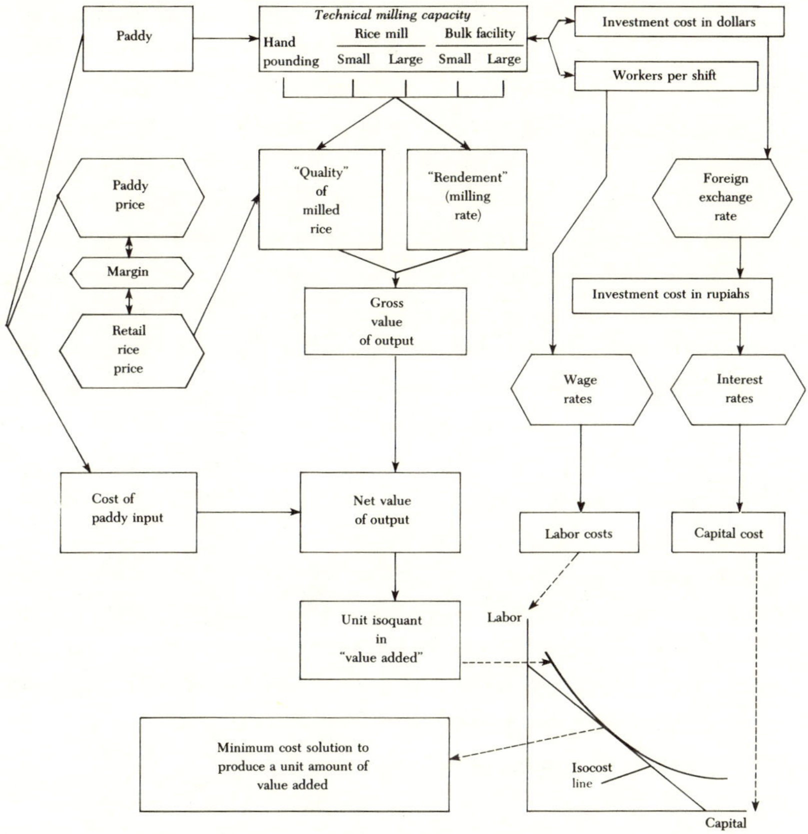

twice as much capital and half as much labor as small rice mills to mill a given quantity of paddy.

The more capital-intensive mills produce more milled rice per unit of paddy input (they are more technically efficient), as well as a better quality of final output, which has a higher price in retail markets. The two factors together determine the net value of output from each milling process for a unit of quantity of paddy input. Subtracting the cost of this paddy yields the net value of output from rice milling. The capital and labor inputs needed to produce a unit amount of this net value—"value added"—then provide the coordinates for a rice milling isoquant which shows the potential substitution of capital for labor while holding value added constant. The relative costs of capital and labor, shown in Figure 3.3 by the isocost line, then determine the least-cost technique with which society can generate value added by milling rice. The factors drawn in hexagons are influenced by government policy, either directly, as with farm and retail price policy, or through policies affecting the macro prices—foreign exchange rate, wage rate, and interest rate. The analysis assumes that a private milling sector makes investments in equipment that minimizes costs. Figure 3.3 shows the information needed and the steps in the calculations to make such a decision to invest.

What is the role of rice price policy in the decision? In the first instance, the absolute level of price for the output from the mill determines the value of "saving" rice through greater technical efficiency. In rice milling, greater technical efficiency can be obtained only from more capital-intensive equipment. Higher prices for rice make higher milling efficiencies more valuable, causing the choice of technique to be more capital intensive and employment to be reduced. The government-determined margin between the farm price and the retail price also directly influences which mills can survive. Hand pounding and small mechanical rice mills, for example, with their low technical efficiency in terms of milled rice output per unit of paddy input, quickly succumb to negative value added as the margin between paddy price and milled rice is narrowed.

The foreign exchange rate is also a key variable in the analysis. Since milling equipment with higher capital intensity has a larger import content, its cost is a direct function of the exchange rate. Devaluation, for instance, raises the cost of capital equipment relative to labor and shifts investment decisions toward more labor-intensive techniques. But the foreign exchange rate also directly affects the price of domestic rice relative to imports. If domestic price policy is influenced by this relative relationship, even in the longer run, offsetting forces will work on choice of technique when the exchange rate changes. The net result is still likely to favor greater labor intensity when devaluation comes, but the effect is clearly quite complicated. Other effects of the foreign exchange rate on price policy are discussed in Chapter 4.

An additional factor might also relate price policy to choice of technique. If one aspect of price policy is to formalize and enforce quality standards in terms of milling outturn, more sophisticated mills with better potential quality control will have a significant advantage. If the price differentials reflecting quality differentials are wide enough, the labor-intensive facilities can be put out of business. Premature quality standards can thus have a negative welfare effect. The difficult task for analysts of price policy is determining when to introduce quality standards and whether to enforce them with price policy.

Marketing systems are characterized by complicated links among sectors and participants, many of them based on trust and fragile expectations. Even subtle changes in policy, especially with respect to prices, can substantially alter the trust and expectations, and ramifications are felt throughout the entire marketing system by all participants—producers, consumers, traders, and the government. Understanding these broader, dynamic effects takes the analyst well beyond the partial-equilibrium framework developed so far. The next task, then, is to expand systematically the horizons of analysis until the full general-equilibrium consequences of a price change are identified. The first step is to determine which border price to use in the analysis, the main topic of Chapter 4. From there, the impact of the foreign exchange rate can be examined. Finally, the analysis in Chapter 4 identifies circumstances under which a change in the commodity's price might have significant feedback effects on both the world price and the foreign exchange rate used to convert it into a domestic price.

4

The Border Price as an Efficiency Standard

If the cost of a commodity in international markets or the potential price for selling it there is to serve as the standard against which the domestic price level is to be evaluated, two questions must be addressed: what is the relevant price in the world market, and what foreign exchange rate should be used to convert the world price into a domestic price equivalent?

The Nature of the World Market Price

Commodity prices are unique in time, place, and form. The price of wheat, for example, can be quoted only for a particular delivery date, location, and specified quality or technical standard. The information that No. 1 Hard Red Winter Wheat sold for $4.18 per bushel in Kansas City in March 1983 is of little use if that wheat was needed in Cairo in March 1983. Even if Egyptian planners want to offer their farmers the equivalent of United States export prices, the March 1983 average price for wheat delivered free on board the vessel (f.o.b.) in U.S. Gulf ports of $4.55 per bushel is not of much help because the planning horizon should be for the coming harvest, not the past one. Moreover, for appraisers of irrigation projects who must judge whether an investment that increases wheat production makes sense, the benefit of Egyptian wheat produced by a project in 1993 may bear little relation even to the cost of about $5.00 per bushel in March 1983 for wheat delivered to Egyptian ports with the costs of wheat, insurance, and freight (c.i.f.) already paid. By July 1985 the same quality of wheat could be imported for less than $4.00 per bushel.

What is the world price of wheat to be used when evaluating Egyptian, Bangladeshi, or Colombian domestic wheat price policy? Apart from the appropriateness of specific varieties and qualities of wheat, a topic not pursued further here, several issues remain significant: the location, terms, and time of the delivery price.

Location

For comparison of domestic prices with world market prices, the relevant location is the country's own border, but domestic farmers must ultimately compete with imports at the point of consumption. For consumers located near major ports, food prices in local markets reflect only small additional distribution charges incurred above the c.i.f. price. At consuming locations in the interior near points of production, commodities produced by domestic farmers are much more competitive with imports.

The impact of location on domestic competitiveness of rice production has been analyzed in West Africa (see Timmer, Falcon, and Pearson, pp. 171–173). When consumption took place at interior locations close to points of production, rice production earned Fr 15.2 per kilogram in social profits relative to a c.i.f. import price of Fr 74 per kilogram at the Senegal port.

Alternatively, when domestically produced rice was consumed in the Senegalese port and production occurred at the most distant interior location, such production incurred social losses of Fr 10.9 per kilogram. The location of the quoted price critically determined whether domestic producers could compete with imports.

Delivery Terms

The direction of a commodity's movement at the border—whether it is being imported or exported—is also important. When wheat is imported, for example, the relevant border price is the c.i.f. quotation. But if exports are involved, the f.o.b. quotation is appropriate. The difference between the two can be quite large. Figure 4.1 provides an illustration based on Zimbabwe corn (maize). As an exporter, Zimbabwe must compete with South Africa and the United States for markets, and so its f.o.b. price reflects the prices for corn delivered by its competitors minus the costs of delivery from Zimbabwe ports. If imports were necessary, however, the relevant c.i.f. price would be significantly higher; it would include the costs of delivery from other exporters' ports to Zimbabwe.

In the early 1980s, Zimbabwe's domestic price policy for corn traditionally was based on export quotations, but it was switched on the presumption that imports of corn would be necessary. The efficiency price for producers was set at the c.i.f. border price, with appropriate differentials for marketing costs to points of consumption. The actual supply curve shown in Figure 4.1, however, shows that the higher producer price reflected by the c.i.f. quotation induced added production and reduced consumption, thus generating an export surplus even larger than that expected at the competitive f.o.b. price. Either the entire export supply had to be subsidized for corn to be

Figure 4.1. Price policy for Zimbabwe maize

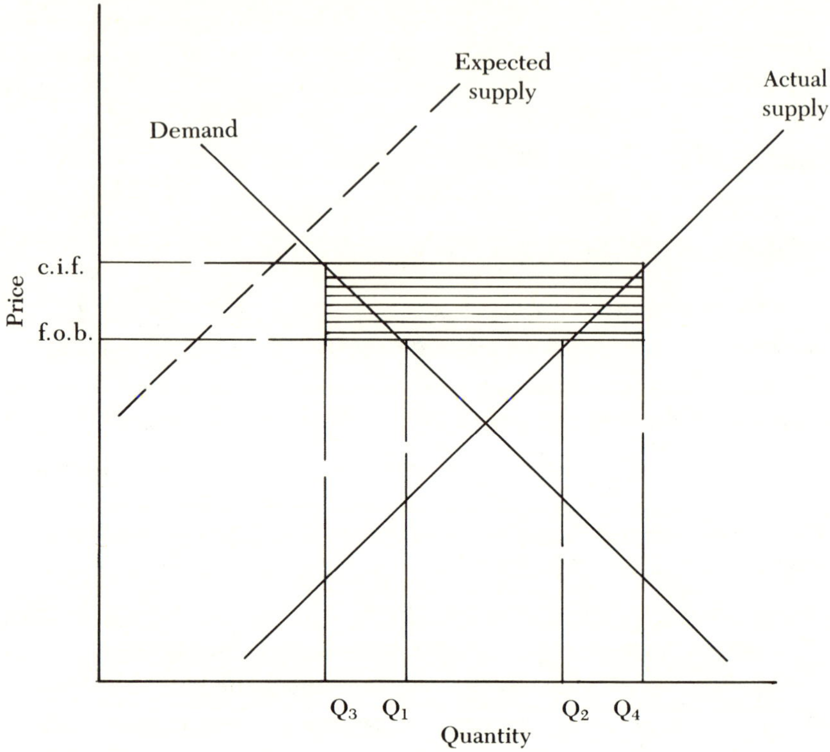

competitive in export markets—at a per-unit subsidy of the c.i.f. price minus the f.o.b. price—or the surplus had to be stored or allowed to spoil, or both.

Perfect knowledge of future production, consumption, and f.o.b. export prices would have led planners to set the domestic currency equivalent price at the f.o.b. level (see Figure 4.1). Production would then equal Q_2, domestic consumption would equal Q_1, and the remainder $Q_2 - Q_1$ would be exported at the f.o.b. price without the need for any subsidies. If farmers are guaranteed the c.i.f. price, however, and domestic consumers have to pay it, production rises to Q_4, consumption falls to Q_3, and the export surplus increases to $Q_4 - Q_3$. This surplus cannot be exported at the f.o.b. price without a subsidy per unit equal to the difference between the c.i.f. price and the f.o.b. price. In Figure 4.1 the total subsidy required is shown by the shaded rectangle. A "mistake" in price policy, implemented with the best of intentions to follow pricing rules that lead to efficient allocation of resources, can turn a competitive export sector into a subsidized burden to the treasury and the economy. At the same time, consumers may be significantly hurt, especially the very poor who must buy their corn in the market, and producers may also be hurt if the government is not able to buy all their surplus production at the "guaranteed" price.[1]

[1]In a comment on an early draft of this essay, Peter Hopcraft offered the following details and observations on the Zimbabwe episode. "The Zimbabwe 1980/81 maize case is clearly one of a 'mistake' in price policy, but the mistake was not merely imperfect predictions by planners. An announced floor price for producers at the export parity level and a ceiling price at the import parity level would have made eminent policy sense, with the final price determined by the size of the crop and the requirements of the market at that price. Instead of that, an announced preplanting price that had been at the export parity level of Z$78 per ton in the previous year, and had resulted in a maize deficit and imports, was nearly doubled to the import parity price of Z$120 per ton. When this price was set, no analysis was undertaken of the likely supply response, so there was no anticipation of the horrendous logistical and budgetary problem created when the combination of the huge price increase and favorable weather generated massive surpluses. (Production rose from 800,000 tons to 1.8 million tons.) The f.o.b. price could certainly not be raised, and it was politically unacceptable to raise the domestic consumer price either, especially in view of the costly surpluses. The basic error was to have a preannounced, politically determined price that was set without reference to supply elasticities and without scope for market adjustment mechanisms. The problem was not merely understandable error in price predictions, it was basic deficiencies in the price determination process. Needless to say, neither the budgetary consequences of dealing with the surplus nor the effects on future prices of large carryover stocks were what the economy needed at that point."

A reviewer made the following comment about this episode: "In view of the experience of India, it is surely worth taking the issue further, because it critically demonstrates the relationship between price policy and structural

Planning price policy when a country might be near the switch point from exports to imports, or vice versa, is very difficult even when the principle of basing domestic prices on short-run border prices is accepted. When a country operates a price policy on the basis of long-run trends rather than short-run quotations, when it is a large factor in the international market itself, and when other important price interventions are also in force, the story is yet more complicated. Figure 4.2 shows an example of all these factors combined in an illustration of recent Indonesian rice and fertilizer price policy (see Timmer, 1985).

In the early 1980s, Indonesia's domestic rice price was above equivalent world market quotations, although most observers felt that the domestic price accurately reflected long-run trends. Fertilizer prices were heavily subsidized, and fertilizer was widely available in rural markets at the low price, roughly one-half the level of equivalent world prices for urea. Indonesia occasionally bought as much as one-quarter of world market supplies in the 1970s, which made it a big factor in a world rice market that is easily influenced by changes in supply or demand from any of its major participants because of the small volume of international trade relative to total world rice production. Indonesia's rice policy and programs had been designed since the late 1960s to achieve self-sufficiency, and the

---

policies. If Zimbabwe had regarded itself under the terms of the Lancaster House constitution as able to undertake a radical egalitarian land reform, there would have been a large rise in the demand for maize by farmers, substantially among the very persons who were also increasing the supply of maize. The 'mistake' would therefore have been to a great extent self-correcting. The problem in India has been very similar—failure to bring about substantial increases in the incomes of the rural poor has meant lack of purchasing power to mop up the extra grain production. Not only price incentives, but also research breakthroughs, can generate this problem—sensible supply policies need sensible (often egalitarian) demand policies if aggregate balances are to be preserved."

Figure 4.2. The world rice market as seen by Indonesia, circa 1984

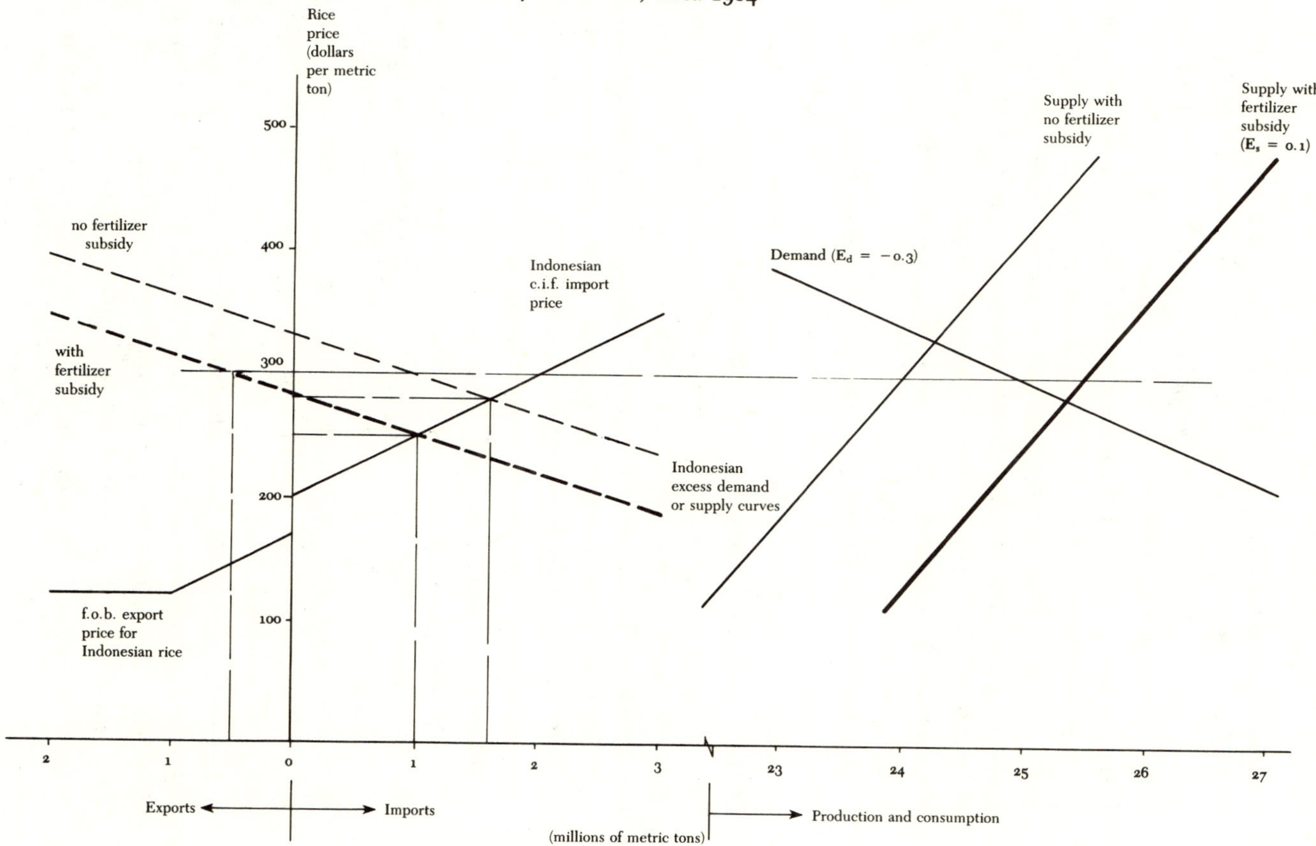

goal was in sight by the early 1980s. Figure 4.2 is roughly calibrated on 1984, and it shows the combined impact of Indonesia's rice and fertilizer price policy on world market prices for rice and on the domestic rice situation.

Indonesia's supply and demand curves on the right side of the figure reflect 1984 levels of production and consumption and are constructed on the basis of short-run elasticities of 0.1 and – 0.3, respectively. The heavy supply curve reflects the prevailing fertilizer prices in the underlying marginal costs of rice production. The light supply curve indicates the effect of removing the fertilizer subsidy.

The left side of Figure 4.2 shows Indonesia's trade in rice. The two dashed lines represent the excess demand for or supply of rice as a function of Indonesia's domestic prices and of the fertilizer subsidy. They are calculated from the difference between supply and demand at each price. The solid line showing import or export price represents recent experience with various levels of Indonesia's rice trade. At a c.i.f. import price, each additional million tons of Indonesian demand from world markets raises world prices about $50 per ton, starting from a level of $200 per ton (c.i.f.) at Indonesian self-sufficiency.

When exports are possible, prices drop by $30 per ton from a c.i.f. to an f.o.b. basis and then continue to decline as exports increase until prices reach $120 per ton, a level at which rice can be fed competitively to livestock. At this point the export price stabilizes. For export quantities of more than 1 million tons, Indonesian rice could sell at approximately constant prices as livestock feed.

Figure 4.2 permits two questions to be addressed. First, holding the fertilizer subsidy constant, what is the impact of a domestic rice price policy that holds urban prices at roughly $300 per ton? At this price domestic supply exceeds demand by 500,000 tons, and prices in world markets fall to about $145 per ton for Indonesia on an f.o.b. export basis. In the absence of a rice price policy (that is, with free trade in rice), domestic and world markets would clear at the same level of about $250 per ton (c.i.f.), and Indonesia would import roughly 1 million tons of rice. Because of the small world rice market, rice price policy converts Indonesia from an importer to an exporter and significantly depresses the price of rice in world markets.

The second question concerns the contribution of the fertilizer price subsidy to these effects. Without the subsidy the free-trade equilibrium price would be about $280 per ton, or $30 per ton higher than with the subsidy, and imports would be more than 1.6 million tons. That is, the *Indonesian* subsidy on fertilizer lowers the *world* price of rice. For this reason, the potential social benefits to the fertilizer subsidy cannot be evaluated simply by reading current world market prices for rice. The benefits to the subsidy include the value of the added rice that is produced domestically because of the subsidy and is no longer imported. The value of these rice imports saved depends on the world price for rice, which is partly determined by the subsidy policy itself. Equally important, the large subsidy on fertilizer has the effect of converting Indonesia from being an importer *even at its current domestic price policy for rice*, with an implied c.i.f. world price of more than $250 per ton, to an exporter at an f.o.b. price of less than $150 per ton. Once again, price policy itself can determine whether a country faces an f.o.b. or a c.i.f. border price.

The potential to switch from importer to exporter status also raises another basic point, which is illustrated in Figure 4.3. Price interventions can be implemented through border policies only if trade can take place across the border, that is, if the commodity is tradable. If domestic price formation takes place within a wide band between f.o.b. and c.i.f. prices, no international trade occurs. Suppose, as in Figure 4.3, that the c.i.f. import price is well above the domestic market-clearing equilibrium price of P_0, while the f.o.b. export price is well below

Figure 4.3. Switching from nontradable to tradable goods

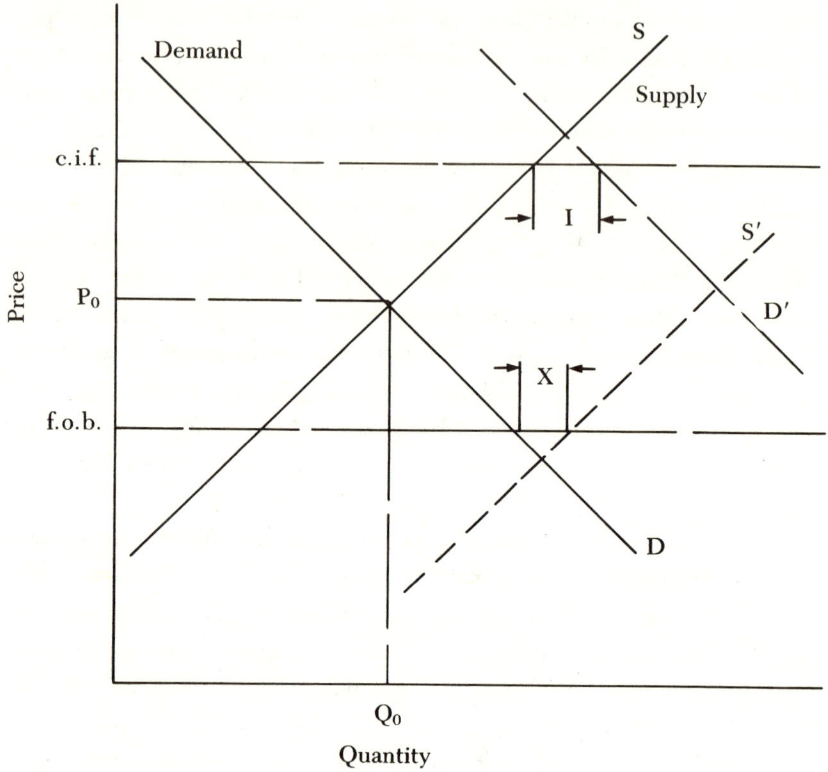

P_0. Such a commodity is a "nontradable" because neither imports nor exports are possible without subsidies. Although analysis of price policies is usually confined to "tradable" commodities that are either imported or exported, Figure 4.3 shows that such a categorization depends entirely on the width of the band between f.o.b. and c.i.f. prices. As internal market conditions shift, or world market prices change, a commodity can easily move from tradable to nontradable status and back. In Figure 4.3, for example, a particularly good harvest that shifts the supply curve from S to S′ would lead to profitable exports of quantity X at the f.o.b. price, whereas growing demand in the face of static supply might shift the demand curve to D′, leading to imports of quantity I at the c.i.f. price. World price levels and transportation costs also play critical roles in determining whether a commodity is tradable at any given time. When freshly cut tulips from New Zealand can be sold in the winter flower markets of Amsterdam, the notion that *any* commodity is inherently nontradable must be questioned.

Still, it is much less difficult for countries to influence domestic prices for commodities that are actively traded across the border than for those commodities whose price formation is entirely determined by domestic supply and demand conditions, as in Figure 4.3. In such circumstances, prices must be influenced by moving the supply or demand schedules, or both, rather than by altering quantities through imports or exports, unless very substantial subsidies are paid. Agricultural research, investments in irrigation and infrastructure, and extension programs can all shift supply schedules. Income growth, family planning programs, and induced changes in tastes can shift demand curves. These long-run strategies, however, are normally considered not as part of price policy but rather as components of overall development strategy. That they clearly influence prices serves as a reminder that the boundaries of price policy and of analysis that seeks to understand its impact are fuzzy indeed.

Time

Once the appropriate location and terms—c.i.f. or f.o.b.—for the border price quotation have been determined, a question that remains concerns the appropriate time horizon for price. Since only spot prices for near-term delivery can be known with certainty, all decisions about the future level of prices entail inherent risk, with the degree of risk becoming larger the farther into the future the price decision must be made. Farmers who plant tree crops such as rubber or coffee obviously face very substantial price uncertainty, but even vegetable farmers with three-month growing seasons can face ruinous price fluctuations between planting and harvest. Three planning horizons for grain markets are discussed here: short-term decisions of whether to supply domestic markets from stocks on hand or to replenish stock levels from imports (or reduced exports); intermediate-term decisions on appropriate prices for producers and consumers when the government is operating a price stabilization program; and long-term decisions about investments to raise output of the commodity concerned.

The short run. The short-term horizon involves day-to-day management decisions about buying and selling in commodity markets. Even when only working stocks are held with no intent to stabilize prices, a company or agency managing grain imports wants to buy from world markets on the best terms possible. Private firms that conduct such transactions must rapidly learn the traders' skills of reading short-run market trends and time their purchases carefully. Otherwise, such firms go bankrupt quickly because commodity trading is an exceedingly competitive and knowledge-intensive business. New information has great market value for those individuals and firms able to process it quickly and change their market positions accordingly.

It is difficult for government agencies to compete successfully in this business, and most parastatal trading companies require at least implicit subsidies to carry on their activities. Even with such subsidies, the profitability record of state-owned enterprises responsible for importing and exporting commodities is quite abysmal, particularly for smaller countries with little expertise in the business. The World Bank estimates that developing countries may pay a premium of as much as 10 percent over least-cost sources of grain simply by not knowing how to manage the purchasing process. Some of this premium is in the form of corruption, of course, but much of it is attributable to lack of expertise. If short-run price stabilization for domestic producers and consumers is important, improving the management skills of importing and exporting agencies is also an important priority for donors and countries alike.

The medium run. The intermediate-term horizon begins to raise important analytical, as opposed to managerial, issues. Most countries value highly the capacity to maintain stable prices for food and agricultural commodities for at least the planning period of farmers—normally a year. If a country wants to provide such stability on a forward planning basis and simultaneously to avoid subsidizing imports or exports with respect to world markets, what price should it choose and what cost must it be willing to pay to guarantee this price?

Most political leaders instinctively assume that price stability is a desirable feature of their economy and devote considerable resources to achieving it. Economists are increasingly doubtful, however, about either the desirability or the long-run feasibility of commodity price stabilization. Newbery and Stiglitz have demonstrated quite persuasively the inherent problems in managing *international* buffer stocks when both supply and demand for the commodity are subject to short-run random shocks as well as long-run competitive pressures from other commodities, producers, and consumers. Domestic price stability is easier to manage because trade can be used as the "balance wheel," unlike international price-stabilization schemes which must rely on buffer stocks to smooth variations in supply and demand.

On the other hand, domestic price-stabilization schemes using international trade create a "tragedy of the commons" problem if all countries attempt to dump their quantitative adjustments to changed real commodity scarcity onto a residual world market. As D. Gale Johnson pointed out in the early 1970s, the resulting world market price becomes increasingly unstable and thus pressures more and more countries to sever the links between their internal prices and world prices (see also Koester). The process is cumulative and leads to progressively worse instability in world prices and progressively higher costs to governments as they attempt to stabilize their domestic prices. In the end, food price stability becomes a luxury only rich countries can afford, as the current record in Europe, Japan, and, to some extent, the United States indicates. A further lesson seems to be that no country has been able to guarantee long-run price stability at "low" prices—below world market trends. All successful stabilization experiences to date have been in countries that maintain their basic agricultural commodity prices above that trend. The possible exceptions to this generalization are the centrally planned countries, but only the Soviet Union has maintained low staple food prices successfully for an extended time. Increasing problems with Soviet agricultural productivity suggest that the policy will be revised before long, as has already happened in China and Hungary, for example.

For countries seeking only medium-term stability, price guidelines up to approximately a year in the future are provided for basic agricultural commodities, except rice, by futures prices quoted by London, New York, Chicago, and Kansas City exchanges. For a modest investment of 10 to 20 percent (the "margin"), it was possible in May 1983 to buy on the Chicago Board of Trade, for example, a contract for delivery of wheat in July 1984. The closing price on a day in late May 1983 for a contract of 5000 bushels was $3.76 per bushel, and this was a decline of $0.055 per bushel from the previous day. The wheat contract for delivery in July 1983 closed at $3.4725 per bushel on the same day, and so the expectation of market participants was that wheat prices in July 1984 would be higher than in July 1983.

Naturally, this does not mean the price of wheat was certain to be $3.76 a bushel in July 1984, nor even that the price in fact would be higher than the price in July 1983. The price had already fallen from the previous day and was down from its high of $4.00 per bushel, and it ultimately reached a high of $4.27 and a low of $3.22 per bushel before the close of trading in the contract. No one can know the market price a year into the future, but the futures market offers buyers and sellers an important option. It allows them to guarantee the price they will pay or receive for a commodity at the delivery time specified. A wheat-importing country might reasonbly have set its price for the 1984 crop year at the domestically delivered equivalent of $3.76 per bushel and known that it would not have to pay a price greater than this for imports if in fact it purchased the needed quantity of contracts at this price.[2]

Do futures markets provide the "right" price policy for the intermediate term? When a government is trying to achieve month-to-month stability of prices but also wants to match the price level to the opportunity costs of imports or exports, should it use the futures quotations as a guide? The evidence on price formation in futures markets suggests that there is little or no bias in these prices; they do not tend to be systematically high or low relative to the eventual spot price for actual delivery. By the same token, however—and this is crucial to whether public sector agencies can use futures markets for

[2]In fact, actual delivery of the commodity specified in a futures contract is seldom made. When a futures contract is about to expire, its price approaches the spot price being quoted for delivery. So long as all types of wheat prices tend to move together, an importing agency can rely on the futures market to guarantee the general price level and do its actual purchasing from the most convenient supplier. The mechanisms by which futures markets can be incorporated into import planning are discussed in Peck.

price policy management—futures market prices are not very good predictors of the eventual price; the futures market price reacts day by day to new information about supply and demand conditions coming to the floor traders at the exchange. A substantial number of individual traders make a living by anticipating this new information or reacting to it within minutes of its availability; however, no econometric models or forecasting agencies, such as the U.S. Department of Agriculture or private economic consulting firms, have consistently "outperformed" the futures markets.

The futures market is thus not a tool to predict prices but rather a tool to manage risk. Contracts to buy or sell a specific quality and quantity of a commodity at a particular date and price permit those individuals, firms, and agencies who place a premium on knowing what price they will pay or receive in the future to transfer the risk of price movement to the anonymous holder of the opposite position for their contract. In principle, a risk premium should be paid for this "service"—Keynes's normal backwardation in commodity price formation. This risk premium should be transferred from producers who are selling forward (hedgers) to users or speculators who are buying. But the empirical record suggests that risk taking is a preferred activity for many futures market traders, and the risk premium is often very small or even nonexistent.

Using futures markets to manage risk requires substantial skills and administrative procedures that permit traders to take future positions that will turn out to be "wrong" roughly half the time. Financial flexibility is required for placing contracts and meeting "margin calls" when prices fluctuate, which involves putting up cash on less than twenty-four hours' notice. For individuals or private firms for which responsibilities can be clearly identified and appropriate incentives provided for good performance, these requirements can usually be met. They are often impossible to meet in public sector enterprises or in parastatal marketing agencies, however. Either highly skilled personnel are not available or they do not serve in any job long enough to become skilled in the specialized procedures of futures training. Financial accounting for public monies is, or ought to be, quite stringent, and thus the capacity of civil servants to speculate or hedge in futures markets is restricted. Administrative planning procedures frequently do not identify import or export movements far enough ahead for futures markets to be of much assistance in providing price stability.

For most government trading agencies in developing countries, the day-to-day trading skills that come from managing the working stocks needed for the implementation of price policy would simply be extended to the intermediate-term planning required for price stabilization within some specified band. This involves reading market price trends, setting domestic prices on this trend, and then having the financial flexibility to subsidize imports or exports if that is necessary to enforce the domestic price guidelines. If the trends are picked correctly, the "good" surprises will about average out with the "bad" over a period of years. Running such a stabilization policy with imports and short-run working stocks is not without cost, however, even in the long run. Price movements are a risk in commodity markets, and usually a cost must be paid to protect domestic producers and consumers from them. These costs include, at a minimum, the administrative budget of the stabilization agency and the opportunity costs of the financial and physical resources tied up in the stabilization effort.

The long run. The third important time horizon for analysis of price policies is the long run—the time when investments in agricultural research, irrigation, and other agricultural infrastructure have their payoffs. When benefit–cost analysis is carried out to evaluate which investments are socially profitable, two important assumptions usually turn out to be critical: whether the benefits will actually begin to flow according to the project's plans (that is, whether there are likely to be delays in construction or implementation); and the price put on the project's output. In considering rice to be produced by an irrigation scheme or cocoa from a land settlement project, knowing the long-run value of the commodity in international markets is crucial to determining whether the project will be socially profitable or not.

Two different sets of actors make decisions in this long-run world, each with different guidelines to pricing. First are the countries whose imports or exports of a commodity have little impact on its price in world markets. For grain, these countries—probably all countries in Africa, all but Argentina, Brazil, and Mexico in Latin America, and all but China, India, Indonesia, Japan, and Thailand in Asia—should perhaps assume that long-run trends of prices in world markets in real terms, after inflation, will continue more or less on their historical path. Figure 4.4 shows these real prices for wheat and corn (maize) from 1866 to 1981. As Martin and Brokken note, the real wheat price shows a significant downward trend over the 115-year period, even after the sharp rise in 1973–1974. Real corn prices do not seem to show any significant trend, although the postwar history of these prices is one of sharp decline.

For "price takers" in world markets, these long-run trends reflect the working out of supply and demand forces over very extended periods. Given the steady, often rapid increases in demand for grain because of growth in population and per capita incomes, the declining real grain prices shown in Figure 4.4 must reflect very substantial long-run supply responses. These responses are a function of technological change, growth in area under cultivation (especially irrigated area), and improved skills of farmers. There are plausible reasons why these factors will continue to generate lower real grain prices. But there are

Figure 4.4. Real prices of wheat and corn, 1866–1981

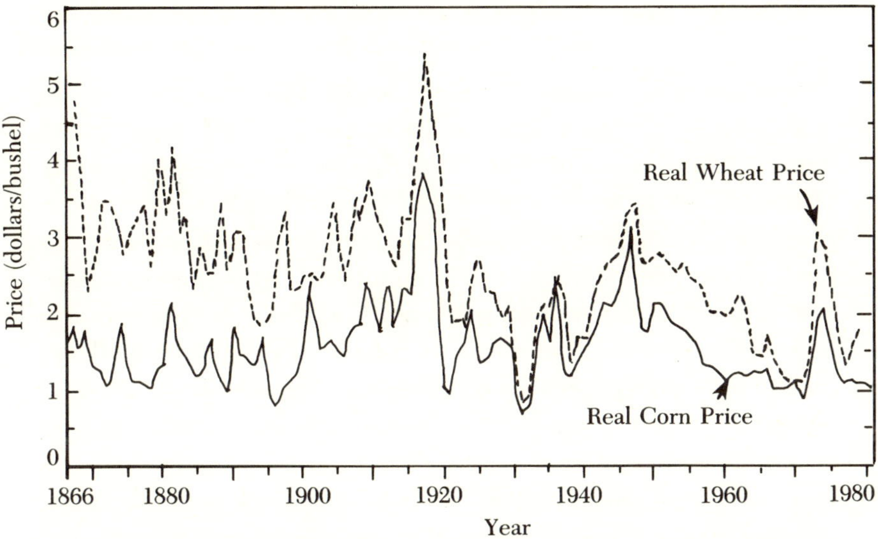

Source: Michael V. Martin and Ray F. Brokken, "The Scarcity Syndrome: A Comment," *American Journal of Agricultural Economics*, 65 (February 1983), pp. 158–59.

also equally plausible ones suggesting that they will fail to do so, and thus higher real grain prices may occur in the decades to come.[3] Although "small" countries clearly have a stake in which direction the world grain economy goes, they have no opportunity to influence it.

[3]A particularly clear-cut example of the "higher price versus lower price debate" appeared with the publication of Simon and Hudson "Global Food Prospects: Good News," and Brown, "Global Food Prospects: Setting the Record Straight." The summaries at the beginning of each article show how wide the differences in expectations can be: "Pessimistic views of the world food picture are rooted in a misreading of historical data. Over the long haul, food production per person is rising, prices are falling, and the outlook is far from catastrophic" (Simon and Hudson, p. 40). "World grain production is losing momentum. Food prices are expected to rise significantly. Moreover, famine returned to the world in the early 1970s after a quarter-century absence" (Brown, p. 48).

The second major group of actors are those countries and international agencies that can influence the long-run price trends because the impact of their production and consumption decisions on international trade is large. Placing a clear plastic ruler over 115 years of data on real prices and then extrapolating the trend line into the future thus results in a major problem for these decision makers. If China and India, for example, presume that long-run prices for wheat on world markets will continue to decline, they might conclude that domestic investments in raising wheat output would not be socially profitable because imports will be cheap. Similarly, if the World Bank decides that long-run prices for rice will be only $250 a ton in 1983 dollars, few proposed irrigation projects or agricultural research investments would satisfy the internal rate-of-return criteria for approval. Precisely because the rice would not be available when those investments would have come on stream, the world market price would be much higher than assumed (especially because the amount of rice traded in the world market is such a small proportion of total world production).

In the absence of some sort of internal consistency check on the part of agricultural planners in the large countries and world agencies, grain prices are likely to exhibit long waves of scarcity and surplus as consistency is enforced by market signals. The consistency requires substantial time and human welfare costs to work out, however, for farmers suffer when prices in world markets fall below even cash costs of production. In the opposite phase, consumers who spend a sizable proportion of their incomes on basic grains are severely affected when prices rise sharply for extended periods of time.

This discussion suggests that two levels of analysis of price policies are required, one for all actors in the short- and intermediate-terms, and one for price takers in world markets in the long run, with an additional consideration added for a relatively small group of actors whose decisions significantly influence long-run prices in world markets. For a country's decisions in the short- and intermediate-terms, probably the best that price policy can do is to use futures market prices as a guide to trends in world markets and to follow these as closely as possible, but not make any effort to "beat" the market's performance. Where futures markets do not exist, as for rice, or where contracts are not available far enough into the future, a simple mechanical rule—a "rule of three"—may be useful. For example, when a decision with a lead time of two months must be made between importing grain or using supplies from working stocks to provide a regular flow to local markets, the relevant prices over the past six months (3 times two months) should be plotted to determine the appropriate "trigger price" for importing.

Similarly, if a price-stabilization policy with a time horizon of one year is being designed for farmers and consumers, the relevant prices (in real terms) in the world market for the past three years should be plotted. The trend of these prices, plotted monthly, can indicate the appropriate border price to use as an efficiency standard. When investment decisions are being made with a payoff ten years in the future, the trend in real prices for the past thirty years should be plotted and extrapolated to the time of the investment's payoff. Although these rules of thumb have no theoretical foundation, they would almost certainly improve the pricing decisions made currently.[4]

For the larger actors whose decisions influence long-run

[4]Actually, Timmer's "clear plastic rule of three" has no empirical foundation either, only a sense that three is about the right multiple. An interesting and potentially important piece of blatantly empirical number crunching would attempt to discover exactly which multiplier would have maximized the probability of being "right" (or more properly, would have minimized the costs of being "wrong" under several plausible loss function rules—minimize expected losses, safety-first, etc.) over the past fifty years or so, for each of the three time horizons for decision making. It is possible that the mulitplier is only 1—the price tomorrow will equal the price today, or that ten years from now will equal the average of the past ten years—or that it is 4 or 7. In Harberger's best tradition of looking at the floor, looking at the ceiling, and picking something in between, 3 seems like a suitable compromise.

prices for important commodities in world markets, a more sophisticated procedure is obviously in order. First, it is necessary to understand the impact of the country's policy on world markets. Even the relatively rough approximation shown in the Indonesian rice example in Figure 4.2 is helpful at this stage in identifying the nature of the domestic–international market interactions. Second, a central information clearing house can offer substantial benefits by keeping track of the actual agricultural investment decisions being made. The Commodity Studies and Projections Division of the World Bank might do this, since the Bank is the largest single investor in agricultural development. Alternatively, the U.S. Department of Agriculture might find this information of substantial competitive interest in light of the important role played by the United States in world commodity markets. In addition, the International Food Policy Research Institute, the Food and Agriculture Organization of the United Nations, or the Consultative Group on International Agricultural Research might perform the task. Indeed, there is much to be said for several organizations undertaking this; the competition would provide incentives to do the best job possible.

Beyond information gathering, iterative models of the major commodity markets are likely to have substantial payoff for an understanding of at least the direction of pressures on long-run prices. Such models are complicated and expensive to build. As more realism is built in, they tend to become "black boxes" in which the endogenous functions are understood by only a very few people who know little about the political and economic realism of the assumptions. Even given this danger, however, the importance of achieving a degree of consistency in long-run decisions among the largest countries and donor agencies justifies further investments in developing more satisfactory models of commodity markets.

Last, the strong tendency of world markets to be self-adjusting must not be forgotten. Long-run supply and demand elasticities for individual commodities are very substantial. On the demand side, the multiplier effect of income-led growth in consumption of grain-fed animal products provides a powerful adjustment mechanism, as does the equally powerful and tragic capacity of high grain prices to force reductions in poor peoples' basic caloric intake. On the supply side, investments in agricultural research for new technology, irrigation, and rural infrastructure are all responsive to price incentives. The lags are quite long, however, from the time the incentives are generated until the higher productivity that has been induced brings real prices back down. The adjustments in demand made during these periods of high real prices have significant consequences for human welfare. By taking a far-sighted view, major donors could help stabilize these cobweb-type cycles. This is not a call for an international buffer stock scheme to smooth out year-to-year fluctuations in prices, but rather for a long-run horizon—ten to twenty years as opposed to two to three years—in planning and coordinating agricultural investments.

Foreign Exchange Rates and Domestic Prices

Even if the world market price for a commodity were perfectly stable, or at least known in advance, a major problem remains for the design of domestic price policy: converting the world price into a domestic currency equivalent. This requires a foreign exchange rate, and there is no completely accepted technique for determining the "correct" exchange rate for this calculation. Even the industrialized countries with foreign exchange rates determined by day-to-day supply and demand forces in international currency markets do not necessarily achieve "equilibrium" rates via this process. In the mid-1980s, for example, the U.S. dollar was widely thought to be overvalued, perhaps by as much as 20 percent relative to its main trading partners, and the Japanese yen seemed to be significantly undervalued. At the same time, all economies are too dynamic and changeable for any optimizing macro model to determine meaningful shadow prices for foreign exchange. The only sensible starting point for making the necessary conversion from foreign to domestic prices is the exchange rate in current use. From this starting point, price policy analysis must investigate the probable direction of bias with respect to relative prices for food and agricultural commodities.

Most developing countries have domestic currencies that are overvalued to a significant degree relative to foreign currency because of high import barriers (either direct controls or tariffs), direct allocation of foreign exchange for "priority" uses, and a strong tendency for domestic inflation rates to be higher than those of their industrialized trading partners. In the absence of controls and trade barriers, a market equilibrium would place a higher price on foreign exchange, thus making imports more expensive in terms of domestic resources while making exports more competitive.

In principle there is no reason why overvaluation should cause a bias against agricultural profitability, but empirically, agricultural commodities tend to be more highly tradable than (protected) industrial goods and services. Most agricultural commodities—grains, beverage crops, fibers—are traded internationally. Those that are not—many root crops, fruits, vegetables, livestock products—are close substitutes for the traded commodities in either consumption or production, and, consequently, their price behavior closely mirrors that of the commodities that are directly traded. For practical purposes, then, the entire food and agricultural sector of a country can be thought of as producing and consuming tradable goods.

If this is the case, the widespread presence of overvalued

Figure 4.5. Impact of foreign exchange rate on domestic prices

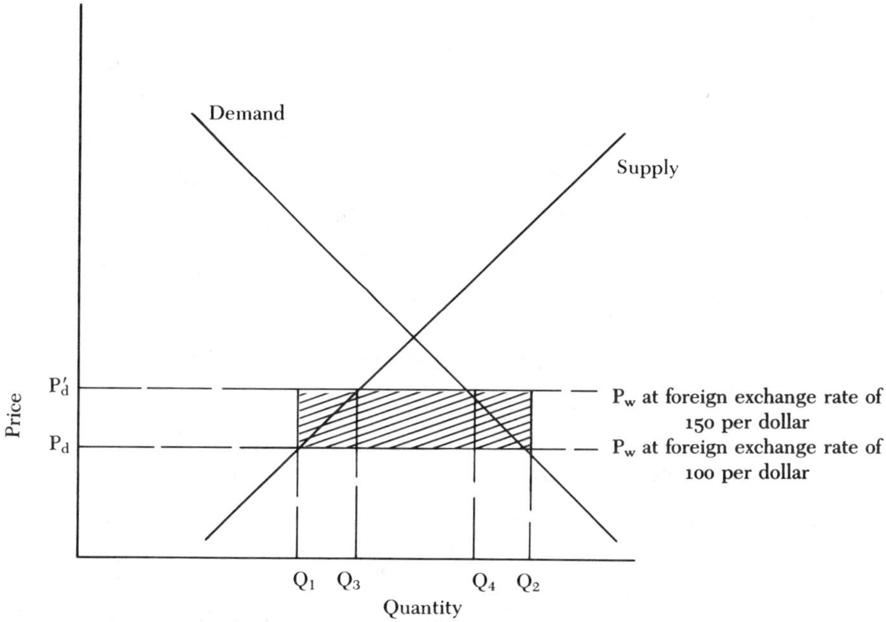

currencies places an equally common burden on agricultural producers—to the benefit of consumers of food and other agricultural commodities (including industry that uses agricultural raw materials). Figure 4.5 illustrates the issues. Suppose as a starting point that the current exchange rate at 100 per dollar underprices foreign currency and that domestic grain prices are equal to world grain prices at this exchange rate. With $P_w = P_d$ and an import gap of $Q_2 - Q_1$, the total import bill in domestic currency is $100P_w(Q_2 - Q_1)$. The cost of these imports is covered by domestic sales receipts, and no apparent subsidy is involved. From the point of view of the agricultural sector, initial appearances suggest that price policy is "right"; it is based on the efficiency criterion of border prices, without income transfers from producers to consumers.

Consider the situation after a 50 percent devaluation, however, from 100 to 150 per dollar (or, if a devaluation does not actually take place, a situation in which the rate of 150 per dollar reflects the realistic opportunity cost of foreign exchange to the economy). If domestic price policy follows the new exchange rate, while world market prices remain constant in dollar terms, the domestic price for commodities rises by 50 percent to P_d'. Because of positive supply response and negative demand response, the quantity of imports is reduced to $Q_4 - Q_3$, and the import bill in domestic currency now becomes $150P_w(Q_4 - Q_3)$. Whether the new import bill in domestic currency is larger or smaller than the initial bill is a function of the supply and demand response, but under the assumption of domestic prices tied to world prices, the entire bill continues to be paid directly by consumers without the need for government subsidy.

Suppose that, in an effort to restrain the inflationary impact of the devaluation, government policy does not permit food prices to rise after the devaluation: the domestic price thus remains at P_d although the world price equivalent is P_d'. If the exchange rate is actually 150 per dollar, P_d can be enforced only with a subsidy equal to $P_d' - P_d$ per ton applied to the original import volume $Q_2 - Q_1$. In this situation the subsidy, represented by the shaded rectangle in Figure 4.5, sets in motion all the consequences for the budget, income distribution, and efficiency already analyzed in Figure 2.4 (see pp. 33–39). At least the budget subsidy itself is quite explicit, however, so policy makers and others are aware that the subsidy exists.

The alternative is for the country to continue to maintain the exchange rate that overvalues its domestic currency at 100 per dollar through import controls, exchange regulations, or multiple exchange rates. The initial situation will then continue to prevail, with the domestic price equal to P_d, imports equal to

$Q_2 - Q_1$, and no apparent budget subsidy. Of course, the overvalued exchange rate *itself* is serving as the consumer subsidy, with precisely the same static welfare consequences for food consumers and producers as if the subsidy were being paid by the budget, but now with broader and more dynamic effects throughout the entire economy.

The foreign exchange rate is the most important macro price to any economy that trades in international markets. It moves the competitiveness of producers of tradable commodities up or down, while it also alters their relative competitiveness because of differences in cost structures, especially in use of tradable inputs whose costs are linked to foreign exchange rates. Analysis of the impact of changed energy prices on agriculture has shown the quantitative significance of this role of exchange rates on the rural–urban terms of trade (see Timmer, 1984). This analysis used changed oil prices for importers and exporters as an "exogenous" shock that has major implications for the macro economy, especially for real foreign exchange rates. The model traced the ripples of these macro shocks through the total economy as well as the agricultural sector. The empirical estimation confirmed the strong impact of both oil prices and foreign exchange rates on the rural–urban terms of trade and on the agricultural sector's productivity performance relative to the rest of the economy. These terms of trade are an important factor influencing short-run profitability of agriculture and rural–urban income distribution. For the Asia–Pacific countries in the sample, a 10 percent devaluation of the domestic currency tended to raise the rural–urban terms of trade by about 1 percent (other trade policy and macro prices held constant). This is quite a powerful effect on the entire agricultural sector in the short run, and individual producers of commodities who use few purchased inputs may feel even more direct influence on their profitability if the price at the farm gate is linked to the export price. In the longer run, dynamic consequences for the sector are likely to be even more substantial and extend to the nonagricultural parts of the economy as well. These broader consequences are discussed in Chapter 6.

5

Disaggregating the Impact of Agricultural Price Policy

Where commodities are fungible and markets are integrated, determining the changes in national aggregates for a commodity's supply and demand is essential for tracing the budget consequences of subsidies or tariff revenues and changes in the quantity of imports or exports when price policy changes. As shown in Figure 2.4, these effects are determined by total supply and demand curves for the country as a whole. Total income transfers from producers to consumers, or vice versa, and total efficiency losses are also depicted in this aggregated analytical framework. Although these supply and demand curves represent national aggregates, they often are used to illustrate the behavioral and welfare consequences for a "representative" producer and consumer. Simply dividing the aggregate relationships by the relevant population number yields the behavior of these "average" decision makers.

Useful as this procedure is—it has been argued in the previous chapters that important insights into income distribution and resource allocation are available from the aggregated analysis—some very important questions of welfare impact are left unresolved if the analysis stops with aggregate results. Welfare effects are felt by individuals or by groups of individuals who are similarly situated. Where differences in initial well-being or in behavior cause the welfare of these groups to be affected differently by price changes, the effects revealed by the aggregate analysis in Chapter 2 need to be disaggregated.

Understanding the impact of price changes on different groups of producers and consumers serves at least two purposes. First, it can identify unanticipated consequences of a policy that call into question its very rationale. A food subsidy intended to improve the nutritional status of the poor that has most of its impact on the urban middle class, for example, would appear to be badly designed. Second, the understanding of disaggregated effects can reveal those who might be most disadvantaged when a policy has to be changed to meet other objectives or external circumstances. A devaluation as a component of an overall reform of macroeconomic policies, for example, is likely to raise food prices and provide greater incentives and income to agricultural producers, many of whom are themselves quite poor. At the same time, however, the higher prices may seriously jeopardize the nutritional well-being of poor urban and landless rural consumers, as well as small farmers who are net purchasers of food. A disaggregated welfare analysis shows this and also identifies possible program approaches to this problem in the short run, while the macro reforms provide a healthier economic base for solving a country's problems of poverty in the long run.

The lives of poor consumers tend to be systematically different from those of consumers who are better off. Their average daily caloric intake is less, the foods they consume tend to be cheaper and less preferred in taste and status, and the sources of their income are often much less regular and secure. When the price of a particular foodstuff changes, the response of the poor is likely to be much different from that of other consumers. In particular, evidence suggests that the poor are much more flexible in substituting one commodity for another as prices change, mostly because this is the only way to maintain adequate caloric intake; the very poor may not have even this flexibility. These subtle and flexible patterns of adjustment are hidden in the usual aggregate demand parameters used for the type of price policy analysis illustrated in Chapter 2. A review of these disaggregated parameters is contained in Waterfield.

Disaggregation can be pursued in many directions and perhaps to an infinite degree. For some purposes, simply dividing the analysis into "rich" and "poor" consumers and "large" and "small" farmers is adequate. The example to be developed here does just that. In some circumstances, further division of income groups is necessary, although it is hard to imagine circumstances in which disaggregation beyond deciles would be useful. Disaggregation of the commodity itself may also be important. Sometimes "food" is a category precise enough to discern the impact of price changes on consumers. But sometimes "wheat," for example, is not sufficiently disaggregated. There may be significant differences in quality of wheat and wheat flour which cause differences in consumption patterns of the poor relative to other consumers. In addition, a wide array of wheat products can be produced, each with its unique consumption patterns. Whole wheat flour may be consumed mainly by the rural poor, for example, and white bread by the urban middle and upper classes. The annual time horizon may need to be broken down into seasons for both producers and consumers. Poor consumers, for example, can be especially vulnerable to high prices just before the harvest, while small farmers may be short of food supplies from the previous harvest. Divisions by age and gender may be necessary if the commodity plays an important role in weaning foods, child nutrition, or provision of additional calories, protein, vitamins, and minerals for pregnant and lactating women.[1]

The appropriate extent of disaggregation depends on the question being asked. Usually the concern is to understand the impact of price changes on the welfare of poor consumers or small farmers because wealthier consumers and larger farmers tend to be well enough off to adjust without tragic consequences. This understanding requires analysis of at least a few household budgets of poor consumers and the decision-making environment of small farmers and landless laborers. Some understanding of urban wage-earner and middle-income household expenditure patterns is also useful for judging macroeconomic consequences of price changes, since these households divert expenditures from luxuries to necessities.[2] Insights might also be gained into sources of potential political opposi-

[1]Sometimes milk or milk powder is thought to be one such commodity, but in most poor countries dairy products tend to be consumed almost entirely by middle- and upper-income households. Pricing policy for milk is likely to have negligible nutritional or welfare impact on poor people as consumers in these circumstances. More worrisome, efforts to keep milk prices low may actually *harm* poor rural producers who depend on one or two dairy cows for their income.

[2]Techniques for obtaining this understanding are provided in Timmer, Falcon, and Pearson. Chapter 2 of that work provides an introduction to the sources of data and the analytical techniques for understanding the decision making of poor consumers. Chapter 3 demonstrates the use of a simple farming system tableau to generate insights into how farm households are likely to respond to changes in their decision-making environment, from possibilities for new irrigation or opportunities for better off-farm jobs to changes in fertilizer or crop prices. The intent of the analysis is to find disaggregated parameters of change for both producers and consumers: income and price elasticities for various food commodities by income class of the household, for example, or crop supply elasticities by size or type of farm. Knowledge of such disaggregated elasticities is essential for the analysis illustrated in Figure 5.1. For evidence of possible general patterns of such elasticities for food consumers, see Timmer (1981); Alderman; and Waterfield.

tion to price changes by studying these groups, even if their nutritional status is not threatened by an increase in food prices.

This chapter does not show how to obtain information on disaggregated decision-making patterns (see Timmer, Falcon, and Pearson); it does, however, demonstrate their role in the welfare analysis of price policy. The simple model analyzed in Chapter 2, a budget subsidy on imported rice, is disaggregated here into "poor" and "rich" consumers and "small" and "large" producers. The differences in their behavior are stylized in the following analysis to keep the discussion reasonably simple, but the differences that are illustrated are based on a reading of the empirical literature.

Figure 5.1 presents the disaggregated analysis in three sections. The aggregate supply and demand analysis shown in the upper graph is identical to Figure 2.4, with two exceptions: the quantity axis has been lengthened to provide clarity when two sets of supply and demand curves are added, and numbers are provided for prices and quantities to illustrate rough magnitudes as well as directions of change. The middle graph shows disaggregated demand patterns for two income classes of consumer, poor and rich. There is no reason to disaggregate supply and demand curves together in pairs, as for poor consumers and small farmers, because the market price is determined for the *aggregate* of supply and demand. Consequently, demand and supply must be disaggregated separately and only after the aggregate analysis has been conducted.

The starting point is a world price P_w equal to $0.100 (per thousand kilocalories), which is translated into a domestic currency equivalent facing producers and consumers at the current (equilibrium) exchange rate. At this price, domestic production Q_1 is 4.58, domestic consumption Q_2 is 5.25, and imports $Q_2 - Q_1$ are 0.67, all measured in millions of thou-

Figure 5.1. Disaggregating the impact of price policy

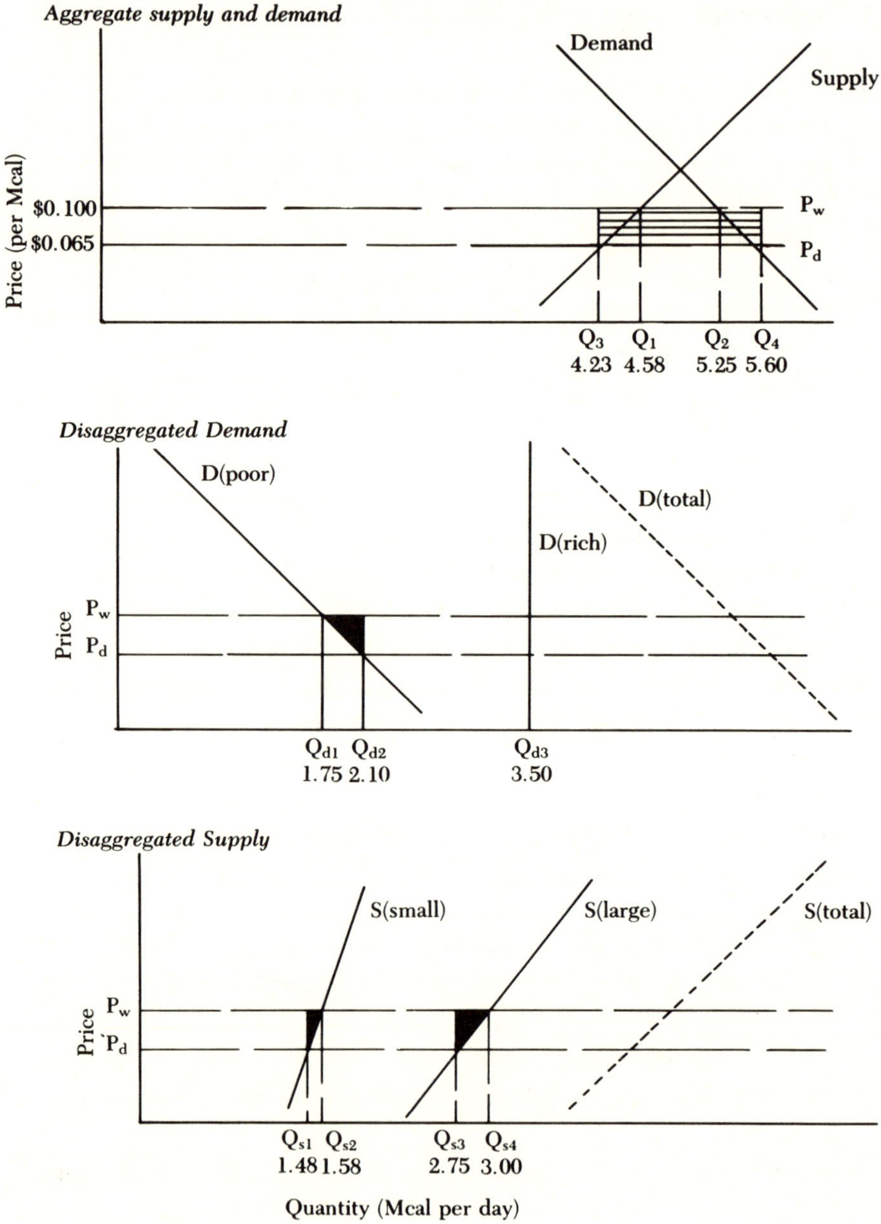

sand kilocalories per day (henceforth, thousand kilocalories per day will be abbreviated as Mcal per day).[3]

A change in price policy is introduced. A budget subsidy on imports is used to lower the domestic price from P_w to P_d. In this example, the change is a 35 percent reduction in the domestic price facing producers and consumers, from $0.100 to $0.065 per Mcal. What is the impact of such a price policy?

At an aggregate level, the lower price causes total demand to increase from 5.25 million to 5.6 million Mcal per day. Since the country has 2 million consumers, average daily caloric intake rises from 2625 to 2800 kilocalories. Both figures are well above the 2500 or so kilocalories per day used as a guideline for average caloric requirements, and little concern for hunger or malnutrition is indicated by these figures.[4] Because of the lower price, domestic production falls, from 4.58 to 4.23 million Mcal per day. Imports rise from 0.67 to 1.37 million Mcal per day, and the budget subsidy required to implement the new price policy is $0.035 per Mcal times this entire import gap, or a total of $17.5 million *per year*.

This budget subsidy is, of course, only a small part of the total resource transfer to consumers. The lower price forces pro-

[3]The issue of quantitative units is not important for the example, and the opportunity is taken to scale quantities into thousands of kilocalories per day (1000 kcal = Mcal per day). If the country's population is 2 million, with 1 million poor consumers and 1 million rich consumers, the quantity axis is measured in millions of Mcal per day. The advantage of this arbitrary scaling is that the impact of price changes on caloric intake can be read directly from the quantity axis. The starting world price of P_w is equal to $0.100 per Mcal, equivalent to about $350 per ton for milled rice.

[4]There is substantial controversy about whether "average caloric (or energy) requirements" have any physiological or social basis. The debate has been carried on both at a biological level and over the use of such requirements for constructing measures of poverty or counting malnourished individuals. A partisan but clear and fair review of the debate is provided by Srinivasan.

ducers to pay an implicit subsidy to consumers of $Q_3(P_w - P_d)$, equal to \$54.0 million per year, plus the loss in producer surplus implied by the triangle equal to ½ $(Q_1 - Q_3)$ $(P_w - P_d)$, or \$2.2 million per year, for a total producer loss of \$56.2 million per year. The total gain in consumer surplus is equal to $Q_4(P_w - P_d)$, or \$71.5 million per year, less the triangle equal to ½ $(Q_4 - Q_2)$ $(P_w - P_d)$, or \$2.2 million per year, for a total gain in consumer surplus of \$69.3 million per year. Note that this total is less than the sum of the budget subsidy and loss in producer surplus (\$17.5 + \$56.2 = \$73.7) by precisely the size of the dead-weight efficiency losses in consumption (\$2.2 million) plus production (\$2.2 million).

The aggregate parameters of change for producers and consumers are also easy to calculate from the upper part of Figure 5.1. Demand rises by 0.35 million Mcal per day because of a \$0.035 per Mcal drop in price; so the (arc) price elasticity of demand is −0.19. Similarly, the aggregate supply elasticity is 0.22. These elasticities are rough approximations and would obviously be somewhat different if calculated on the basis of a price change from \$0.065 to \$0.100 per Mcal, rather than in the other direction. The general order of magnitude, however, is satisfactory for the purposes here.

This is the ending point for neoclassical price analysis of market aggregates. The border price and foreign exchange rate are each taken as given. No ramifications on the rest of the economy or the dynamic implications on patterns of production and consumption have yet been considered (as they are in Chapters 6 and 7), and all producers and consumers are presumed to be adequately represented by the average. What happens when the impact of the price change is disaggregated?

Impact on Consumers

The middle graph of Figure 5.1 shows two different patterns of consumption for this commodity, which for simplicity is taken to represent total food consumption. Rich consumers, represented by demand curve D(rich), show no response at all to price changes. Their price elasticity of demand for this commodity is zero, and they consume 3500 kilocalories per day no matter what. Poor consumers, however, represented by D(poor), do not enjoy such luxury. Their poverty leads them to adjust their food intake when prices change, partly for obvious real income effects and partly because of pure substitution effects. In the example shown, poor consumers initally buy 1750 kilocalories per day when prices are at world levels, $0.100 per Mcal, but this amount increases to 2100 kilocalories when prices are subsidized to a level of $0.065 per Mcal. The price elasticity of demand for poor consumers implied by this change is −0.57, which is three times the average price elasticity. The gain in consumer surplus for the poor is $24.6 million per year, although the entire efficiency loss in consumption is borne by the poor.

The nutritional gains from the significantly larger daily caloric intake might be a more appropriate measure of the welfare effects of the price subsidy. No matter what the average caloric "requirements" for this country are, an increase from 1750 to 2100 kilocalories per day for half the population would have a salutary effect on overall health and nutritional status and probably on work productivity as well. The effect is overstated in this example because no allowance is made for reduction in intake of some commodities that have become relatively more expensive in the face of the price drop for the commodity analyzed. Such commodity substitution when *relative* prices change can substantially alter the partial-equilibrium results for a single commodity whose price has changed.

It can also be seen that although the consumption levels of the rich do not change at all, rich consumers gain more of the total income transfer induced by the subsidy policy than do poor consumers—$44.7 million per year for the rich as opposed to $24.6 million per year for the poor. (The total is the same as the aggregate gain in consumer surplus measured in the upper graph in Figure 5.1, $69.3 million per year.) The larger gains to the rich occur for an obvious reason: they consume more of the commodity than do the poor, and the subsidy applies equally per unit of commodity rather than per unit of "need." This pattern occurs for any commodity with a positive *income* elasticity of demand, and the higher the income elasticity, the stronger the pattern. Nearly all the subsidy for milk products, meat, or gasoline, for example, is captured by upper-income consumers, whereas subsidies for cassava, millet, or low qualities of wheat or rice tend to have a more equal distribution of benefits and often a concentration on the poor. Even average income elasticities provide a clear picture of who gains or loses from a price subsidy or tax, and income elasticities by income class permit very precise estimates of the short-run incidence of these interventions into price formation.

Different patterns of consumption by rich and poor consumers do raise the potential for some subsidies to be effectively self-targeted by the commodity itself. Coarse grains and root crops, for example, are consumed primarily by poor people, whereas wheat and rice tend to have positive income elasticities well into the upper-income ranges. Commodities with low or even negative income elasticities—dried cassava is a classic example—can provide vehicles for subsidies that are reaped only by those households willing to consume the commodity (so long as consumption of the commodity by livestock is somehow prevented). Searching for these commodities and analyzing the impact of subsidizing them—on consumers, the food system, and the rest of the economy—is a main topic in the book by Timmer, Falcon, and Pearson.

The short-run consequences for consumption of a subsidy are not the whole story, however. Subsidies for consumers can have devastating spillover effects even when the commodities are consumed only by the poor. It makes little sense, for instance, to lower the *market* price of a commodity consumed primarily by the poor if the commodity is produced by very poor farmers. Not only does the lower price cause problems of market supply, but, since poor farmers are the major producers, rural welfare problems at least as significant as those facing the beneficiary consumers can result if prices are driven down by the subsidy program. When dried cassava, for example, is subsidized as a way of targeting food subsidies directly to poor consumers, the farmers who sell dried cassava grown on their small upland plots can have their incomes dramatically reduced if the market price is driven down.

One possible solution to this dilemma is to provide a purchase price that is higher to farmers than the retail price is to consumers. But the potential then to "recycle" the commodity—for consumers to buy cheap and sell dear to the government at the higher farm price support level—can seriously jeopardize the financial integrity of the program. Such recycling problems can be avoided if the commodity purchased from farmers is clearly distinguishable from that sold to consumers. This is easy for rice, where paddy (unhusked rice) serves as an appropriate commodity for farm price supports while milled rice can be sold to consumers. (The logistics agency must then find a way to mill the paddy.) It is somewhat harder for dried cassava, but some sort of nutritious cassava-based gruel, perhaps made by adding vegetable oil and peanut meal (for caloric density and protein) and hot peppers (so the chickens won't eat it) to an otherwise starchy base, could serve to separate the farm commodity from the consumer's. Where such a potential exists, separate price policies for producers and consumers might be implemented if the spillover effects on the rest of the food system were not too disruptive. Understanding the prospects and problems of such policies is thus an important task for food policy analysts.

Impact on Producers

Comparative evidence on the supply response of small farmers compared with that of large ones is not yet extensive enough to provide even a rough basis for generalization. The lower graph in Figure 5.1 is constructed around a common stereotype of farmers in developing countries: small farmers produce primarily for their family's subsistence needs and are not so responsive to market prices as larger farmers who have more surplus production for sale. As with many stereotypes, there may be some basis for the image, but the diversity probably swamps the differences. The figure shows small farmers producing 1.58 million Mcal per day at P_w and 1.48 million Mcal per day if the price falls to P_d, a short-run supply elasticity of 0.18. Large farmers show a more substantial supply response, with an implied elasticity of 0.24. Their base production before the price subsidy policy is 3.00 million Mcal per day. Small and large farmers together produce a total of 4.58 million Mcal per day, the total shown in the upper graph of Figure 5.1.

When prices are pushed down from P_w to P_d, large farmers reduce their production along the supply curve S(large) from 3.00 to 2.75 million Mcal per day. In conjunction with small farmers' reduced supply to 1.48 million Mcal per day, aggregate supplies fall to 4.23 million Mcal per day. Since aggregate demand has risen because of the price drop, additional imports are needed to supply the market at the new subsidized price of $0.065 per Mcal. Total imports are now 1.37 million Mcal per day, costing $50 million per year, of which $32.5 million is paid by consumers and $17.5 million is paid directly as a subsidy from the government treasury.

As already noted, this budget subsidy to consumers is only a small share of the total subsidy they receive. Farmers pay the rest through reduced gross revenues and lower incomes (at least implicitly). Because large farmers are producing nearly twice as much as small farmers in this example, they bear nearly twice the burden of the income transfer. Their loss in producer surplus is equal to the new production level (2.75 million Mcal per day) times the drop in price ($0.035 per Mcal, or $35.1 million per year) plus the loss in profits because of reduced output ($1.6 million per year). Total losses by large farmers are $36.7 million. Small farmers' losses are $19.6 million per year. As a consistency check, it should be noted that the sum of small and large farmers' losses in producer surplus, $19.6 plus $36.7 million, is equal to the $56.3 million loss per year from the aggregate analysis.

This is the point at which the joint production–consumption decisions of farm households become especially critical. One approach is to assume that the gains in consumer welfare to the farm household just cancel losses in producer welfare. The small-farm sector then just nets out of the aggregate analysis. Evidence suggests, however, that small-farm households make substitution choices in consumption in a fashion similar to other poor consumers, as well as responding to production incentives. This implies substantial arbitraging in commodity markets, perhaps by growing high-value rice for sale in order to purchase and consume lower-value corn or cassava. If this behavior is representative, the small farmers' supply response to price may be even more than that shown in Figure 5.1.

The disaggregated results illustrated here present a difficult dilemma for policy makers because the consequences of the price subsidy for income distribution are quite mixed. Rich consumers gain more than poor consumers, but the nutritional status of poor consumers is significantly improved nonetheless. Large farmers are taxed implicitly in larger measure than small farmers, but if small farmers carefully make joint production–consumption decisions, the lower price for a high-value commodity may dramatically lower their real incomes and consumption possibilities. The tax incidence of the budget revenues needed to implement the price subsidy policy has not been analyzed here, but it is easy to imagine widely different patterns, from the incidence on poor producers (taxes generated from export crops raised by small farmers on marginal land, as, for example, rubber in Indonesia) or on rich, foreign consumers of copper or gasoline (taxes generated from exports of minerals or oil).[5]

These consequences for income distribution are only for the short run. The changed output levels in the rural areas are likely to change the demand for hired labor (and perhaps the supply as well). The lower rural incomes will have secondary effects through expenditure patterns and their impact on demand for labor. The changed circumstances in rural markets and the availability of subsidized food in the city may affect household migration decisions. These and other roundabout and dynamic effects of price changes on income distribution and the rest of the economy are discussed in Chapters 6 and 7, but they raise a final point about the value of disaggregated price policy analysis for understanding the political economy of price changes.

If the policy debate were whether to eliminate the subsidy and raise the price to world levels, the short-run consequences for income distribution would appear quite perverse. Large farmers would reap nearly $2 out of every $3 in additional income generated by the higher prices—and almost all of the additional real income if consumer losses in the households of small farmers roughly equal their income gains. Wealthier consumers would lose more in total real income than would poor consumers, but, unlike the poor, they would suffer no nutritional consequences. As a result of the rise in food prices and the decline in consumption by the poor who buy most of their food from markets, significant hunger and malnutrition might

[5]A good survey of the literature on tax incidence relevant to developing countries can be found in Gillis et al.

appear. Wealthier consumers, however, tend to be the urban middle class—civil servants, workers in large-scale industry, students, the military—whose political voice would be heard clearly when their pocketbook was squeezed, even if there were no nutritional consequences to the pinch.

Once subsidies are introduced or are in existence, such consequences would be quite evident in the short run. The perceived impact of eliminating subsidies can then be completely immobilizing when governments attempt to remove them, even though the aim is to achieve a more efficient allocation of resources in order to build a long-run productivity base capable of supporting the welfare goals achieved, but simultaneously threatened, by the subsidy policy. The nature of this dilemma and the policy tensions emanating from it are the subject of Chapter 7.

6

Beyond Single-Market Analysis

All markets are linked by substitution possibilities in production and consumption. Sometimes individual markets are isolated, segmented from other markets for the same commodity or factor by transportation costs, institutional barriers, or government regulation and activity. But even when rural labor or credit markets seem fully insulated from urban or national economic forces, indirect economic activities such as migration or remittances provide roundabout and subtle linkages.

At the level of principle, all analysis of agricultural price changes should be set in a full general-equilibrium context, in which the market outcome for any one good or service ultimately depends on the activities in all other factor and product markets. As a guide to understanding the impact of policy, however, the pervasiveness of such linkages is immobilizing. Do we really need to know the general-equilibrium consequences of a price change for tomatoes in the Philippines? Or for corn in Indonesia? Or for rice in Thailand? It all depends. The aim of this chapter is to explain on what it depends.

Micro to Macro: From Single Markets to General Equilibrium

For a micro economist analyzing individual market behavior, "macro" means general equilibrium, or the simple (but consistent) sum of all micro market outcomes. To a macro economist, on the other hand, "macro" tends to mean aggregate employment and gross national product. The micro economist is interested in the vector of prices that clears all markets. The macro economist is interested in the monetary and fiscal policies and effects contained in the aggregate outcome. It is not surprising that researchers in these two subdisciplines have had a difficult time communicating with each other and agreeing on models that are consistent with each other's view of the world. Much of modern economic theory and some empirical effort has been devoted to bridging this gap. Rational Expectations models, for example, can be crudely described as efforts to build aggregate economic models that track basic macroeconomic data but retain utility-maximizing decision makers at the micro level.

Such efforts are important to analysts trying to understand the impact of commodity price changes beyond an individual market because the models seek to identify precisely the type of linkages that create spillover effects. The goal is not necessarily to build an elaborate model of these effects—although at times that may be necessary and desirable—but rather to establish a frame of mind for identifying important linkages and judging their importance before a major effort at modeling is made.

The approach here is to work from simple to complex, starting with the single-market analysis outlined in Chapter 2. When the model is interpreted carefully and thoughtfully, it provides clear directions on where to look for the most immediate impact on other markets. It requires a willingness to ask how the process of adjustment is undertaken. Exactly what does it mean for a supply curve to have a positive slope? What are farmers doing? A reduction in rice production is likely to be accompanied by diminished use of inputs and by some increased effort on other crops or activities, including greater willingness to enter rural labor markets. Quite immediately, then, at least three other markets—input markets such as for fertilizer, alternative output markets such as for cotton, sugar, or corn, and rural labor markets—are identified which have potentially significant links to the rice market and would adjust to the rice price subsidy discussed in Chapter 2.

These multi-market effects are not easy to calculate even once identified because of the likely spillovers from the initial market intervention. Unlike single-market analysis, there is no simple graphical framework, and researchers are forced into numerical calculations even as a rough first step. A more satisfactory approach is to build a multi-market model with as many interconnections as are necessary or within the reach of available data. Such models are much simpler to build if analysts start from current market environments and calculate approximate changes through incremental adjustment coefficients and partial elasticities than if they attempt to find globally consistent functional forms for each market's supply and demand curves (see Braverman, Ahn, and Hammer).

Even fairly simple multi-market models can yield important differences from single-market analysis. Table 6.1 shows an example from the World Bank's work on agricultural pricing in Malawi. A 14 percent reduction in the price of corn (maize), the country's staple food crop, was analyzed. With just the partial-equilibrium results, corn production declined by 6 percent and average farmer income fell by K 29 million. The multi-market results showed the same decline in corn production, but a switch in resources, especially farm labor, led to increased tobacco and peanut (groundnut) production by 2 and 3 percent, respectively. Because these crops are exported, foreign exchange earnings increased by K 7.5 million. The lower corn price also caused fertilizer use to drop by 15 percent. After the adjustments included in the multi-market model were accounted for, farmer incomes dropped by only K 18 million instead of the single-market estimate of K 29 million.

Table 6.1 shows that an important feature of a multi-market

Table 6.1. Comparison of single-market and multi-market analyses for Malawi

Commodity	Price	Changes in production and income in response to 14 percent reduction in farm-gate price of maize (annual): Single-market analysis	Multi-market analysis
Maize	Subsidized	−6.0%	−6.0%
Tobacco	Taxed	0.0	2.0
Peanuts	Taxed	0.0	3.0
Fertilizer	Subsidized	0.0	−15.0
Smallholder average incomes (nominal)		−K 29 million	−K 18 million
Foreign exchange		0.0	K 7.5 million

Note: All four commodities were traded in markets with government-fixed prices and adjustments in quantities used to clear the markets.
Source of data: Kirchner, Singh, and Squire.

model is an understanding of how the respective markets operate, especially whether they are endogenously price flexible or whether all adjustments are through quantities, as in the Malawi case. Multi-market models, in fact, are applied primarily in those cases in which governments enforce fixed prices in several markets and so must plan for variable quantities if price stabilization is to be successful.

Endogenous price determination for regional or national markets further complicates the analysis. In combination with alternative crop competition for regional or national resources, such as land, water, or labor, this complication requires a sector-wide model to trace the important effects of price policy changes (see Kutcher, Meeraus, and O'Mara). Agricultural sector models have been used since the mid-1960s to check the consistency of agricultural development plans and to examine the resource requirements of alternative cropping patterns. With the advent of much faster solution algorithms for the optimization segments of the models, efforts have recently been made to incorporate realistic pricing factors into the models, including rising supply curves, falling demand curves, and endogenously determined market prices for all commodities in the model, on the basis of maximizing the sum of consumer and producer surplus in the agricultural sector.

Unlike single-market or multi-market equilibrium models, sector models contain detailed farm models as the basis for supply response and alternative crop choices in the face of changed prices. In principle, similar detail for differently situated consuming households could also be incorporated, and normative own- and cross-price elasticities in both production and consumption could thus be built into the model. Calibrating such normative elasticities against actual market experience is an important but difficult task for sectoral modelers. It is precisely because econometric evidence on cross-price elasticities is so shaky that farm- and household-level optimizing models are built into the sectoral model in the first place.

Just as a price change for a single commodity can spill over into other markets and as multi-market effects can have important sectoral consequences, so too can agricultural sector adjustments have significant macroeconomic consequences. Two broad categories of sectoral spillovers require at least a rough effort to model macro consequences:

when budgetary revenues and expenditures are endogenous to the policies being analyzed but are allocated across sectors, as when large food subsidies that draw on general tax revenues are used to implement a price policy;

when macro prices are endogenous to the problem on a national level and cannot be understood on the basis of sectoral market segmentation.

Rural labor or credit markets that are only weakly linked to urban or national markets can usually be analyzed with sectoral models. But when wages in national labor markets, foreign exchange rates in currency markets, and interest rates in capital markets for savings and investment resources are affected by a change in a commodity's price, a macroeconomic model that enforces market clearing and general equilibrium is essential for consistent results.

Computable general-equilibrium models are not yet capable of capturing many of the key linkages between individual commodity prices and macroeconomic outcomes. Rapid progress is being made on the technology of these models, however; already the capacity to build and solve such a model substantially exceeds our understanding of the important connections that must be built into the models if they are to provide useful insights. It is possible to speculate on the nature of those connections and some likely macro consequences of price changes for important food staples, for example, but not to offer results of revealing empirical analysis in this area.

Even here the insights are limited. Food and agricultural policy is seldom considered as part of macroeconomic policy, or vice versa. Planning agencies, central banks, and ministries of finance are responsible for macroeconomic policy, while food agencies and ministries of agriculture design farm and retail price policies. The macro agencies have an input into these deliberations (frequently a veto) because of the consequences of food and agricultural policies for the government budget, but the reverse is seldom true. Few ministers of agriculture contribute to the macro policy debate.

The Macroeconomic Impact of Agricultural Price Changes

In many poor countries with large rural sectors, or simply with a large fraction of personal income spent on basic food commodities, the macroeconomic consequences of food price changes are quite significant. The obvious impact is through the size of the government budget devoted to price subsidies, but the effect on foreign exchange might also be quite important when price changes alter the volume of imports or exports. Countries with major industrial sectors producing behind tariff barriers for domestic consumption can also experience fluctuations in employment, through mechanisms first identified by Keynes, when food prices change. Higher grain prices, for example, force consumers to increase the budget share devoted to grain and to reduce purchases of other goods and services supplied by domestic industry and workers. If the reduced domestic demand cannot be compensated for by increased exports and money wages are not flexible downward, widespread urban unemployment and reduced national income can result. A similar impact can also be felt in the very short run in rural areas if demand is diverted away from labor-intensive services, such as carpentry, and from fruits, vegetables, and dairy products, which have a significantly higher rural labor content than the basic grain, especially if it is imported.

In the other direction, sharply lower food prices might depress demand in rural areas for many goods and services that are produced domestically as the income effect on farmers of lower grain prices begins to be reflected in their consumption patterns. Lower prices for staple grain stimulate demand in urban areas for manufactured consumer goods and imports. Either way, the consequences of price changes of basic food and agricultural commodities for the budget, employment, level of prices, balance of payments, and income distribution can be substantial. Macroeconomic consequences are not the same as general-equilibrium consequences, however. Lower food prices may benefit the poor for only a short while, for example, if nominal wages adjust to lower real costs of hiring labor and supplies of unskilled labor are relatively responsive to real wages.

The Budget, Deficits, and Inflation

Although dynamic effects of price changes are discussed in Chapter 7, these merge imperceptibly into long-run macroeconomic effects, and keeping the two completely separate is virtually impossible. To be sure, there are some immediate, short-run macro effects of a price change. The subsidy introduced as a price policy change that has been used as an example throughout this essay has such an immediate impact: the budget suddenly incurs a cost of $17.5 million per year, $8.75 per capita, or perhaps 2 to 3 percent of national income in this make-believe society.

Depending on how the budget cost is financed, a variety of macro effects are possible. If subsidies are shifted from hospitals, for example, to food imports, the direct consequences for the macro economy via the budget may be nil. If the subsidies are financed by deficit spending and money creation, significant inflation may result. Since inflation does not influence all prices equally (indeed, food prices can be held stable via imports), this means of financing the budget subsidy could have substantial effects on income distribution. Owners of physical assets may prosper, holders of money assets may lose, and unskilled laborers may find their wages lagging behind rises in living costs.

The inflation may also have an unforeseen effect on the foreign exchange rate. As domestic prices rise relative to those of foreign trading partners, a fixed exchange rate causes the domestic currency to become progressively more overvalued, thus increasing the implicit subsidy on imports, including food imports. If budget subsidies tend to lead to inflationary pressures, a dynamic and potentially cumulative process has been set in motion. This inflationary process and the budget deficits that cause it are among the first elements of fiscal reform a negotiating team from the International Monetary Fund will look for when a country needs standby assistance because of balance-of-payments problems. Whatever the long-run contribution of such a fiscal reform to efficient economic growth, and it is likely to be quite positive, the short-run political and human consequences of sharply reduced food subsidies are often explosive. In such situations the relationship between effective and feasible macro policy and the need to disaggregate price policy analysis is quite clear.

The direct effect on the foreign exchange rate is seen when the initial subsidy causes imports to rise. This alone may put some pressure on the foreign exchange rate, but if the subsidy is financed through budget deficits that result in inflation (or "more" inflation), additional pressures will be felt on the exchange rate. If the rate remains fixed, the currency becomes progressively more overvalued, thus lowering the real costs of imports and increasing the volume of food imports. The demand curve for the subsidized food commodity will shift out as other prices become relatively higher. The supply curve will shift back as other supply opportunities become more profitable for the country's producers. Consequently, food price policy involves more than movements *along* the commodity's supply and demand curves. When multi-market and macro effects are taken into account, it is clear that the price policy can also affect the *location* of the curves themselves. These feedback effects from subsidized imports mean that, unless checked by devaluation or reduction of budget deficits through cuts in other sectors, the domestic food sector can become *progressively* more dependent on imported food. Incomes of producers and their productivity suffer, and consumers may well become vulnerable to shocks in world grain markets in the long run as domestic production declines.

The rise in food consumption per capita associated with this situation cannot long be sustained unless outside financial resources are available—from oil revenues, for example, or foreign aid. At the same time, job creation is slowed under the impact of the overvalued domestic currency, and the incomes of the poor may fall. If this happens, the distribution of the gains in consumption because of the budget subsidy will be more skewed than that shown by the aggregate results, and even the short-run nutritional gains will be jeopardized.

None of the consequences of price policy for the budget and inflation is discernible from numerical calculations based on reasonably simple models. Too many accommodating forces and variables are also at work for such effects to be identified. Substantial mechanical and methodological progress has been made since the mid-1970s in the construction of computable general-equilibrium models. These models are very complicated and yet remain highly restrictive in their ability to capture the subtle long-run effects of price policy changes on income distribution and the structure of the economy. One example of the potential of these models is provided by de Melo and Robinson; a general review is in Robinson.

Labor Markets

Dimensions of macro effects other than through fiscal effects on inflation, however, may be more quantifiable. Taylor has constructed several simple macro models that capture some of the Keynesian consequences of food price changes. These models are most applicable in an economy with a substantial industrial sector operating behind tariff barriers that permit it to be of high cost, and hence uncompetitive in export markets. The fairly rigid quantitative demand for food and its large share in the average consumer's budget mean that food price changes lead to substantial reallocations of budget expenditures to food (when prices increase) or away from food (when prices drop). As a consequence, the shift in consumer spending on industrial goods can have a major impact on total demand and utilization of industrial capacity, thus affecting urban industrial employment. Many developing countries do not have industrial sectors of sufficient size for this to be an issue, but the newly industrialized countries of Latin America and perhaps India and Egypt fit the description.[1]

Trade orientation is an important element of Taylor's models. Import-substituting countries operating behind protective tariff barriers tend not to be able to find additional export markets if domestic demand falls when food prices rise. Countries using export-led growth as the engine of development, such as Taiwan and South Korea, are less susceptible to such shifts in domestic demand. They pay a price, however, by being dependent on the vicissitudes of *external* demand for their industrial products. As the world recession in 1982–1983 demonstrated, this demand does not ride a one-way "up" escalator.

Very little is known about the impact of agricultural price changes on rural labor markets. This is partly because, as Binswanger and Rosenzweig document and analyze, rural labor markets are extremely heterogeneous. Such markets depend not only on local supply and demand conditions but also on local institutions and traditional hiring arrangements. Rural labor markets are frequently linked to credit markets, access to land, and even commodity markets by individual and personal ties between employers and laborers. These linked market arrangements vastly complicate the analysis of a commodity price change as it ripples through these various markets. Fairly simple supply and demand models with rural and urban labor markets connected by migration may give useful insights, such as those that emerged from the study of rice price policy in Thailand (see O'Mara with Vinh), but even these models are highly country-specific.

[1]For the macro impact of food prices, see especially Chapter 5 of Taylor, which contains the model called "price policy and the food people consume."

Structure of the Economy

It is important to understand the two-way street between price policy and the macro economy. For some countries the food economy and the macro economy are more or less the same thing. These countries are sufficiently poor that most of the gross domestic product is generated from rural-based production. When processing and marketing activities are added to the purely agricultural base, the point is simply made stronger. Good crops, at stable prices, mean years of good incomes. Crop failures plunge the entire economy into depression.

Few economies remain so heavily dependent on agriculture, although the entire food and agricultural sector seldom declines below 20 percent of total economic activity even in rich countries. Yet for countries with only 20 to 30 percent of gross domestic product stemming directly from agriculture, and another 10 to 20 percent from input processing and marketing activities, the linkages between rural and urban sectors can remain quite strong. Research at the International Food Policy Research Institute (IFPRI) shows that multipliers of employment from changes in rural incomes can be substantial, provided consumer prices do not favor imported goods over ones that are produced domestically. Demand for fruits, vegetables, and livestock products is quite elastic with respect to both rural and urban incomes, and these products can be very labor-intensive if produced appropriately from domestic resources. Similarly, the demand for light industrial goods—pots and pans, bricks, tiles, clothing, dishes, glass, bicycles, and furniture—can provide the basis for domestic industries that are very employment intensive.

Although the IFPRI studies focus on technological change in agriculture, price policy provides an equally powerful policy lever for generating the rural incomes that create this kind of demand. Several countries—for example, Japan, South Korea, and Taiwan—have consciously used price policy to protect their agricultural sectors, at least partially in order to generate more rural purchasing power and a better consumer base for their industrial products. A common alternative has been to drain incomes out of rural areas by explicit or implicit taxes on basic agricultural commodities and subsidies for consumers. Such rural taxation may be inevitable in the very early stages of development, but it quickly has diminishing returns as an urban economy develops. Such urban-biased price policies raise the disposable incomes of consumers, especially middle- and upper-income urban consumers (unless the subsidies are targeted directly to the poor). These consumers tend to have different expenditure patterns from those of their rural cousins. Urban consumers buy more imported goods and products of the urban industrial sector, which tends to be more capital intensive than rural industry, especially if heavily protected. Again, these secondary effects on employment and income distribution that are generated outside the agricultural sector are not easy to quantify, but investigation in specific circumstances can be revealing (see Krishna; Mellor; and Hazell and Roëll).

Investment

The overall level of investment in an economy is a key variable in even the simplest macro models. The potential of price policy to change the level of investment, primarily because it changes income distribution between producers and consumers and between urban and rural areas, is thus an important macro concern. The composition of the investment is not necessarily a macro issue, but it does strongly affect the dynamic impact of price policy on income distribution, long-run growth, and the structure of the economy.

At any given time, savings rates tend to rise with household incomes because low-income consumers must spend most of their resources on current needs. Upper-income households save more in the form of financial assets and make much larger investments in their family's human capital in the form of expenditures on education and health care. A price policy that transfers incomes to poor consumers at the expense of upper-income consumers can be expected to lower the aggregate savings rate for the society in the short run, and vice versa. Although a low volume of savings used very efficiently in investments can allow a country to outperform in growth terms another country with a high savings rate but very low efficiency of investment, it remains obviously true that a high savings rate *and* high investment efficiency lead to the best performance of all. Current empirical literature on the impact of positive real interest rates on both volume of savings and efficiency of investment suggests that the two are correlated. A food price policy that avoids economy-wide subsidies to consumption, either explicitly or through an overvalued domestic currency, generates additional savings. These savings, in turn, have greater impact on growth in productivity in the context of a macro price policy package that contains "efficiency" levels of interest rates (and foreign exchange rates).[2] Consequently, a potential synergy exists between pricing consumption goods to reflect border prices and charging positive real interest rates to reflect the social opportunity costs of capital.

The sectoral investment impact of an agricultural price policy that provides additional financial incentives to producers is also likely to reinforce the macro effects. Farm households have higher savings rates than those of households where income is earned entirely from wage labor, mostly because of the opportunities producers have to make productive investments di-

[2]For further elaboration of these issues, see Timmer, Falcon, and Pearson, Chapter 5, as well as Timmer (1983).

rectly in the farm enterprise. These investments might be very short term—fertilizer, pesticides, or seeds, for example—but they can also have significant longer-term effects. Careful development and maintenance of irrigation works, land leveling, long-run enhancement of soil fertility, and investment in higher-value tree crops or livestock operations all have an impact on productivity in the long run. They are also quite responsive to the price environment facing farm households. One of the major tragedies attending a sustained bias against farm-level price incentives is the loss of this source of potential growth in agricultural productivity.

One last impact of agricultural prices on investment must be noted. Even without accepting the entire induced-innovation hypothesis of Hayami and Ruttan, it is clear that government investments in agricultural research, infrastructure, and other supporting services for rural areas are very much a function of the perceived scarcity of food. Prices are one loud and clear signal about the degree of this scarcity. Domestic investment decisions that are based on social benefit–cost analysis need not, in principle, incorporate actual market prices into the analysis. Since government is doing the investment, social prices, rather than private prices, can be used to value benefits and costs.

There is clearly a tension, however, between an investment decision based on one social value for an agricultural or food commodity and a domestic price policy that places a different social value on it, if only because of the confusion engendered in interagency coordinating committees. The different time frames of implementation—importing to achieve price stability in the short run as opposed to making investments for productivity growth in the long run—mean that full consistency is not absolutely essential. But the analytical and managerial complications of actually implementing these differences between short-run and long-run planning prices are serious indeed. Perhaps more important, private investment decisions, particularly at the farm level, are made on the basis of expected private prices, not the planning prices used in social benefit–cost analysis. Forgoing this source of growth must be included as one of the potential costs of a price policy that provides general food subsidies to all consumers.

The Limits of Price Policy Analysis

Because of the linkages between food price policy and the rest of the economy, price policy analysis is necessarily complex. Nearly all government policies have an impact on the structure of incentives facing producers and consumers, and their subsequent decisions influence the entire macro economy from all directions. One response of analysts to such complexity and pervasiveness of agricultural and food price policy is to argue that market adjustments would take care of everything automatically if governments simply let the markets operate and find their own appropriate prices. So long as factor and product markets are capable of allocating resources reasonably effectively, government policy makers need not worry about all the indirect, roundabout, and dynamic effects of price policy. Despite the desirability of such a market approach as a long-run strategy, there are three substantial problems in the short run with this approach:

It presumes that all domestic prices can be brought into line with international opportunity costs at more or less the same time;

It presumes that domestic product and factor markets function well enough to satisfy the neoclassical conditions for efficient resource allocation. This presumption may not hold without government investments in roads, marketing facilities, and communications and information systems. In addition, there may be rural labor markets in heavily populated poor countries where the "market" wage leaves substantial numbers of underemployed and unemployed workers;

The short-run consequences for income distribution, including the nutritional effects, of the market solution might not be acceptable to the society.

All governments intervene in the formation of domestic food and agricultural prices. While the free-market solution provides a powerful paradigm for understanding the impact of these interventions, its normative judgments are not sufficiently robust, or realistic, to provide satisfactory policy answers. Its power lies not in providing an optimal answer but in identifying alternative costs and benefits of different interventions. The search for the full range of effects of price policy interventions, no matter how far-ranging or complicated, is legitimate and important. Policy analysis will always be a process that is at least as intuitive as it is quantitative.[3] To treat it exclusively as one or the other is to miss both the strength of the underlying economic models of resource allocation for illuminating important social choices and the necessity for society, rather than analysis, to make those choices.

Perhaps what emerges, most persistently, from observing the formal analysis of a country's price policy and the resulting policy decisions is the constant need to make decisions, even decisions to do nothing. Many countries maintain price policies that are patently not conducive to their long-run economic growth, despite rather sophisticated analyses that illuminate the consequences of the policies. Simple inertia may be partly to blame: bureaucrats receive less criticism if something goes

[3]The point is made in an especially compelling way in the introductory chapter to Johnston and Clark.

wrong from current policy than from a visibly new policy. The anti-growth policies may also be attributed to the immediately tangible political costs of a change in policy relative to the tenuous and widespread dispersion of the anticipated economic benefits. But often the reluctance to implement a more effective price policy stems from a failure of coordination between analysis and policy. Analysts may solve problems that policy makers are not interested in; policy makers may be unwilling to think about the potential scope of analysis or may focus solely on its limitations.

Effective analysis can contribute to breaking all three of these bottlenecks to policy reform. Clarifying benefits of a change in policy and making them vivid for a wide audience can help overcome inertia. Where negative consequences for the welfare of the poor come with a change in policy, analysts can identify vehicles for targeted subsidies that can ease the transition. Where political reactions from upper-income groups are feared, phasing of policy changes may be effective in defusing the opposition, as may direct and honest explanations of the need for change. And analysts can make an effort to solve the problems policy makers face—not the ones that are easy, convenient, or intellectually appealing to solve.

7

Markets, Politics, and the Dynamics of Price Changes

Throughout this essay has run a provocative but almost subliminal message: that markets provide the appropriate context in which to discuss the efficacy and efficiency of government price policy interventions. A different perspective might also have been taken—one that has in fact been adopted by many developing countries. The paradigm of central planning with quantitative targets and allocations as the decision variables, with prices set to reflect social values and needs, has been offered as an alternative to the market perspective on price policy interventions.

The planning paradigm has considerable theoretical and emotional appeal. In a review of an edited volume on comparative economic systems, Montias discusses the reasons for this appeal as expressed by the theoretician David Conn:

> Conn writes that the centrally planned economy "offers great promise for control and coordination, [which] is in part a desideratum because it could lead to improved performance." This stems from (1) "a capacity for a superior to recognize externalities"; (2) the ability of planners to assess the preferences of the

> system's members and act on these "rather than give people votes on the marketplace in proportion to their income or wealth"; (3) the ability to allow politics to supplement or supplant economics in decisionmaking; (4) the capacity to consider the preferences of future generations. Now he has defined a CPE [centrally planned economy] as an economy where there is a central agency which has the capacity to (a) collect information from other members of the economy; (b) define the preferences of the economy as a whole; (c) determine the course of action that is to be pursued by the economy; (d) implement this plan; and (e) actually exercise its authority (Montias, pp. 627–628).

Although it is hard not to be taken aback by the chasm between this description of central planning and its actual operation anywhere in the world, the appeal of the paradigm is obvious. Central planning, in principle, allows society to make important allocative decisions on the basis of human needs rather than according to scarcity values of prices determined by the purchasing power of a highly unequal world market. Simply because substitution of wheat for corn in world markets means the demand of American, Japanese, and European consumers for steak makes wheat expensive at times is no reason why Mozambique should also make wheat expensive for its own poor consumers. The quantities of grain needed to fulfill a society's goals can be part of the state plan and can be produced directly or imported. In this way, social priorities, including the elimination of hunger, can be met well before an economy has grown sufficiently wealthy for all consumers to be adequately nourished without government interventions on their behalf.[1]

[1]The necessity for a food stamp program in the United States, however, suggests that market-oriented economies may never become rich enough for all consumers to be able to afford adequate diets from their own earned incomes. Even rich countries have hungry people if there are no food interventions.

The planning paradigm, which owes more to Fabian socialist thought than to Marxist–Leninist theory, is used as a basis for economic organization for a majority of the world's population. The extent of control over planning varies widely, however, from the fairly rigid quantitative targets for nearly all sectors and enterprises in the Soviet Union to the indicative, but not prescriptive, central planning done by such nominally socialist countries as India or Indonesia. Within this vast heterogeneity of centrally planned economies is an emerging uniformity of result: central planning with direct quantitative controls places a heavy burden on the performance of the food and agricultural system. But at the same time, food remains the most basic of human needs, and one that all planned economies try to meet as a high priority. Out of the tension between the failure of planning to generate a modern, efficient agriculture and the urgent desire to see that all people are adequately fed has come a great deal of experimentation in socialist countries with respect to the organization of food systems. This has come at the same time that market-oriented economies have sought efficient interventions to improve the access of their poor people to food.[2] No single solution to this food policy dilemma is likely to emerge as the answer for all societies, but the underlying importance of markets as a key to all the solutions is being recognized.

Markets can perform three functions. They can provide the

[2]For a fascinating description and analysis of the experiments with "production responsibility systems" in China, see Hartford (1983). The communes are being dismantled as economic decision-making units, and production decision making has been returned to individuals, households, or very small groups of households. Markets have been reestablished in rural areas to provide both incentives for greater production and signals to planners about surpluses and scarcities.

For a series of reports on government efforts to introduce food interventions in market-oriented systems, see several of the research reports of the International Food Policy Research Institute.

physical and institutional setting for the transformation of agricultural commodities in time, place, and form. As noted in Chapter 3, these technical marketing activities must be carried out in all food systems, for they involve storing the commodity from harvest to time of consumption, transporting it from farm to city, and processing it into forms desired by consumers. Not all food systems carry out these activities cheaply and efficiently, but they all perform them.

The second potential role of markets is to provide mechanisms and institutions to transfer ownership (or at least physical possession) of commodities. Shops, stalls, village marketplaces, or supermarkets provide consumers access to food if they have money. The farmer's field or house, a roadside turnout where a truck can stop, or wholesale markets near a warehouse or mill offer farmers the opportunity to sell. Only fully subsistence farmers do not need markets to play this role. Everyone else must trade something of value, frequently their labor, for the opportunity to receive food and other goods and services.

The third potential function of markets is price discovery. When physical transfer of ownership takes place, a bargain has been struck. The buyer parts with the cash equivalent of the price, and the seller parts with the commodity. All societies see some of this sort of exchange—buyers and sellers agreeing on a price for a commodity or service—but in many societies, such exchange is either illegal or restricted to minor commodities. Basic foods are nearly always important in such societies, and attempts are made to enforce centrally directed terms for the exchange—the price for the commodity. The buyers and sellers are not free to alter these terms, and no matter whether supplies are scarce or in surplus, the price of exchange (and sometimes the quantity as well) is set by the state.

The appeal of such a policy if it is carried out reasonably effectively is twofold. First, stability in basic food prices is assured, and this is enormously popular with urban residents who must purchase all their food from modest wage income. Experience in Japan and Western Europe has shown that even high stable food prices are popular so long as they convey a powerful sense of food security. Price stability at the farm level is also popular, for with it farmers do not have to develop complicated strategies for dealing with risks from both markets and weather; they can devote their talents to the day-to-day tasks of ensuring the best crops possible given the changing physical environment for farming. The main issue, of course, is stability at what price. Poor consumers need low and stable prices, while farmers want high and stable prices. The second advantage of a central government's control over crop marketing and price formation is the potential to set the two prices separately. The subsidies needed to pay the marketing costs incurred between the farm and consumer can come straight from the treasury as a reflection of social priorities. The costs of this approach can be surprisingly high, as the model pictured in Figure 3.2 suggests.

This approach to price policy is incompatible with the use of markets for price discovery through the interaction of many buyers and sellers. In these circumstances, markets play the first two of their potential roles, but price discovery as a market function is truncated by the planning guidelines. This can happen not only in socialist countries but also in market-oriented countries that are trying to enforce specific price levels for particular commodities. Price policy interventions that rely on quantitative controls thus inevitably dampen, and sometimes destroy, the role of markets in price discovery.

Price policy need not have this devastating effect on markets. As explained in Chapter 3, an entire structure of prices for a commodity is generated in an efficient economy even when floor and ceiling prices are implemented. This structure has dimensions of space, time, and form and depends on competitive interaction among buyers and sellers in local markets as they respond to their own transportation, storage, and processing costs. Allowing this process of price discovery to go on within local markets does not jeopardize the overall goals of maintaining the price band at a socially desired level. Freely functioning markets and active price policy are not incompatible.

Markets play both an engineering and an economic function. The first two roles—the storage, transportation, and processing function and the transfer of ownership—can be thought of as engineering functions. Technical efficiency in their design and operation is crucial to low-cost performance. In planned economies, no economic analysis at all may go into investments in these dimensions of food and agricultural marketing. Whatever equipment and system the engineers designate as efficient is put into the plan. Actual investments are then made according to priorities in the planning agency.[3]

The economic function performed by markets occurs when the costs of transportation, storage, and processing are reflected in the willingness of sellers actually to transfer the commodity to buyers and when the total cost of the purchase influences the consumer's decision of whether or not to buy. Price formation itself thus takes place in markets as a simultaneous concomitant of the costs to society of producing (or importing) and consuming (or exporting) the commodity.

This price discovery is important because of the information contained in the prices. When prices are formed in reasonably competitive markets, the scarcity of the commodity as measured by the marginal costs of production (or supply) is matched by the willingness of consumers to substitute this commodity for something else in their expenditure bundle. Prices are signals. When they are high, the commodity is

[3]For an example of an engineering approach to planning marketing investments at the national level, see Timmer in Timmer et al. (1975, Chap. 1). For the engineering approach at the firm level, see Wells.

scarce. Producers are stimulated to expand output, and consumers are encouraged to restrict purchases. When prices are low, they signal producers that only minimal effort should be expended in production, and they induce consumers to expand their use of these cheap goods and services. Both high prices and low prices are important; they provide signals to decision makers to adjust their behavior to reflect a disequilibrium between long-run supply and demand.

The full importance of markets to the agricultural sector now begins to become apparent. Food production is inherently unstable, and even demand is often hard to foresee, especially demand in export markets. More than perhaps any other major part of the economy, a country's food system tends to be moving from one disequilibrium setting to another. What adjustment mechanisms are available to cope with this continuous disequilibrium and all the risks and costs associated with it? How can government intervention make the adjustment mechanisms more flexible?

Market systems have literally millions of decision makers on both the supply and demand sides of the disequilibrium responding through individual adjustments and incentives to bring the system back into balance. None of these adjustments is easy, and the process is often made more difficult by government efforts to partition markets into administratively controlled zones. Low prices can force small farmers to sell their land. High prices can force poor consumers to restrict their food intake. Although these adjustments are potentially painful, they are more or less automatic in market economies if governments let them happen. Programs to foster market adjustments by lowering transportation costs, providing better access to capital, and speeding the transfer of resources from one crop to another or from one sector to another can reduce the pain.[4]

[4]A reviewer made the following observations about the trade orientation implied in the previous paragraph: "This is the classic 'bias towards trade and

Centrally planned food systems have no such automatic adjustments. Indeed, the planners themselves may not even know about the disequilibrium if all price information is flowing in only one direction, from the planning office down (see Hartford, 1984). It often takes long lines at food shops or signs of urban discontent to break through the bureaucratic consciousness so that information about scarcities and surpluses arrives at the planning office. Such information may not be very helpful, however, if it is too late for planners to make marketing or pro-

---

marketing.' Why is it economically efficient, or a proper role for the State, to encourage the shift of private resources away from local towards long-distance uses, via the subsidization of transport and marketing? Even more important, why should a financially strapped State divert scarce resources from productive investments towards marketing investment in this way? Why should not one wait for the development of substantial surpluses to generate the necessary (economic and political) pressures for better marketing infrastructure—and meanwhile use scarce State power to strengthen these pressures by improving production resources?"

In a separate comment, the same reviewer made the following observations: "By providing free transport (road systems), subsidized communications, free or subsidized market information, etc., governments subsidize trade (and distant trade more than local trade) and therefore implicitly discourage production for subsistence or even for local or nearby consumption and exchange. There is nothing in the 'gains from trade' arguments, which are of course perfectly sound under their own assumptions, to justify the subsidization of trading and marketing even under those assumptions. Yet it has become the new orthodoxy that 'free roads are fine, free food's a crime.' It needs to be stated firmly that there is no economic logic in this pro-marketing bias, which is in process of converting excessive parts of African budgets, in particular, to infrastructure which encourages exchange between points of minimal consumption, rather than increasing output so that the existence of a surplus, rather than the existence of pressure-groups and articulate politicians, leads to the creation of marketing infrastructure such as roads."

While infrastructure investments obviously should be evaluated on the basis of their social profitability, it is somewhat difficult to see how substantial surpluses can be generated in remote areas without roads and markets if the private sector is expected to move high-yielding inputs and new technology into these areas. The government can do it, of course, at high expense and substantial probability that the technology will be inappropriate and the input supplies delivered late.

duction reallocations. The dynamic adjustment of a food system to new costs and opportunities is hobbled when they are hidden from the decision makers who must ultimately make the adjustments.

One of the great paradoxes of modern economic systems is apparent here. The agricultural and food sector is one of the most "public" in terms of policy and program needs, but at the same time it is one of the most "private" in terms of day-to-day decision making in production, marketing, and consumption. The paradox is explained by the role of markets and the myriad of decisions that are coordinated by those markets. It is the well-known instances of market failure—when markets do not provide adequate supplies of public goods or do not perform efficiently—that gives agriculture its public dimension: the need for government policy to ensure competitive standards, along with investment in agricultural research, roads and marketing infrastructure, irrigation, communication facilities, grades and standards, and even price stability, which may not be provided by competitive private markets to an extent that is socially adequate. The government's policies and public investments determine the efficiency and dynamism of a country's agriculture more than in almost any other sector. But at the same time, millions of individual households make the day-to-day decisions that actually generate the efficiency and dynamism.

The sources of market failure in developing countries are many, as Meier notes:

> Market failure can exist in various forms: (1) the market does not function properly—the case of market imperfections; (2) the market result is incorrect—the case of externalities; (3) no market exists for the relevant activity—the case of public goods; and (4) the market yields undesirable results in terms of objectives other than resource allocation. In these cases of market failure,

market prices do not exist, or do not reflect the true value, or are irrelevant.

A state of underdevelopment is in large measure synonymous with pervasive market imperfections. A variety of market imperfections can reduce the efficiency of private market performance. The properties of efficiency in resource allocation and Pareto optimality of distribution, which characterize a competitive equilibrium, depend on the existence of a competitive set of markets. But an underdeveloped economy is to some extent an "empty economy" with an incomplete set of markets. Moreover, unlike the perfectly competitive markets we have previously postulated, a market may be deficient in the provision of information or subject to lags in adjustment, or characterized by insufficient competition among firms.

If information is inadequate or inaccurate, market signalling will not be read or correctly interpreted by agents in the market. A certain resource allocation will occur, but it will differ from the efficient allocation based on adequate and accurate information. It can then be argued that it is in the public interest to have the government correct these informational deficiencies. The scope for the provision of public information is wide in a developing economy—from agricultural extension services to technical assistance to information on world market opportunities.

Time lags also impede the efficient working of the market. While resources might respond to market signals, they may do so only after an unduly long delay. To accelerate the supply response to a rise in price, for example, the government may want to undertake policies to stimulate supply. This is especially relevant for overcoming factor immobility, via training and educational measures and other policies designed to reduce barriers to entry in an occupation or industry. After acquiring information about a market opportunity, the economic agent must be willing and able to react on this information. To fulfill these conditions more speedily, the government in a developing country may have to undertake policies to overcome the fragmentation or segmentation of markets and improve the operation of markets. And yet while these market imperfections may prompt government

> intervention, it is necessary to weigh the costs of arbitrary direct administrative controls and to be sure that they do not exacerbate an already fragmented economy (Meier, pp. 137–38).

Significant tension exists over the widespread failure of markets in the food and agricultural sectors of rich and poor countries and the desirability, even necessity, to use markets as the vehicle both to reach individual producers and consumers efficiently and to generate an evolving flow of information about costs and returns in the sector. Joseph Schumpeter wrote in *Capitalism, Socialism, and Democracy* that capitalism and free-market economies would decline in democratic countries, to be replaced by socialism and political controls, because most people cannot understand the complex mechanisms by which the benefits of market systems are generated. This complexity causes political support to wane for the contractual property rights essential for market economies to operate efficiently. Without a clear consensus supporting the often skewed results of market outcomes, special economic groups use their political skills to intervene in markets for their own financial advantage, a process that gradually undermines the allocative role of prices. One important task for economists is to explain the consequences of this process to a wider public.

Schumpeter's vision of the changing nature of the fundamental institutions that define the basic contractual arrangements of our societies calls in question attempts by economists and economic historians to model the long-run changes in the structure and performance of national economies in terms of price and income variables. Precisely because the rigor of these models holds in abeyance the changing "rules of the game" and ideologies of participants, they cannot, alone, capture the complicated evolution of modern economic systems as they respond to exogenous shocks or internal policy changes.

It is not just economic modelers who would like to assume

away the complexity of interdependent economic systems. Policy makers and populations at large also seek simple and understandable answers to the difficult problems of modern economic life. Ever since the debate over repeal of the English Corn Laws, economists have argued that simple answers frequently make for bad policy and that what is good politics in the short run is bad economics in the long run. T. W. Schultz's comment about some economists being "'yes men' in the halls of political economy" shows that the debate is still lively. Moreover, contrary to Schumpeter's expectations, the growth in government regulation and control over the economies of democratic societies that he so accurately foresaw, and of stronger socialist organization in the totalitarian societies, has not been achieved with minimal losses in economic performance.

Two fundamental approaches have been used to resolve the tension between the failure of markets and the need to achieve efficient resource allocation. One has been to displace markets from their economic functions and to rely on planning allocations or direct parastatal controls instead. The more-tightly regulated and internally oriented economies, however, have suffered declining growth rates, widespread shortages of consumer goods, and open dissatisfaction of the population. Ironically, the reasons have to do with the complexity of economic systems, the extraordinary amount of information communicated through prices, and the multiple decisions involving capacities for processing vast amounts of information required of all successful economies in a globally interdependent world. Most of those decisions simply cannot be made efficiently or effectively in a central planning agency or regulatory control board.

The use of quantitative controls has become progressively less efficient in a highly dynamic world where opportunities and potential losses are a function of information flows. As a consequence, new planning experiments have been tried.

These have taken several forms: the Hungarian orientation to agricultural exports led by decentralized cooperatives and firms reacting to market signals; the Chinese experiment with production responsibility systems and rural markets; and an expanded role in several Latin American and African countries for private traders in legal markets as competition for government marketing parastatals, and the withdrawal of parastatals from activities that require direct interaction with farmers and consumers, with a new focus on wholesalers and intermediaries or export markets (see Austin).

The alternative approach has been to use government interventions to strengthen the efficiency of markets within a food policy framework. It is a surprise to many that the agricultural sector of a developing country is an intimate component of a complex economic system, linked to it through markets and price signals. A vision of largely self-sufficient peasants outside the money economy and the incentives signaled by changing prices recalls a simpler and nostalgic era, but an era passed a century ago in the United States and now nearly gone for most of the subsistence farmers in developing countries as well. For better or for worse, agricultural and marketing systems in developing countries are now an integral part, often even a dominant part, of their countries' market economies.

This market approach incorporates investment in public goods and a commitment to solving the problems of poverty and hunger through targeted interventions that reach the poor without seriously distorting the signals generated in the markets. Investments in a marketing system can lower marketing costs directly, thus easing the food price dilemma of providing low prices to consumers but sufficiently high prices to producers, and these investments can also increase the efficiency and competitiveness of the marketing system by lowering barriers to entry.

Changes in food and agricultural prices, in price policy, affect the structure of the economy, and in return the agricultural economy is influenced in roundabout and dynamic ways. It is difficult for analysts of price policy to capture these dynamic effects. If high food prices induce technical change, for instance, static analysis may provide seriously misleading estimates of the actual welfare transfers between producers and consumers and the ultimate impact on the government budget. By changing the opportunity costs of holding land for small farmers relative to large ones, a government can induce quite unknowingly, through price policy, changes in land tenure patterns or in choice of production technology. The distribution of income can change not just between producers and consumers but also within rural areas through indirect effects on hiring patterns and rural wages. These effects on income might then have long-run consequences for nutrition and health not captured by the static analysis (Behrman and Deolalikar). Because some outcomes reflect the difference between life and death, actually measuring these consequences raises serious methodological problems.

The dynamic effects of food price policy interventions are likely to dominate the static effects, but the main issue is whether they reinforce or cancel each other. Much remains to be learned in this area, and comparative case studies of modern economic history rather than econometric analysis are likely to provide much of that knowledge. These studies require a political economy framework that focuses on the role of prices in the process of structural change and incorporates information on institutional response, creation and reactions of effective political lobbies, and the changing attitudes of society about the fairness and functioning of the economy (see Reich, Endo, and Timmer; Anderson and Hayami).

The historic evolution of price policy from discrimination against farmers in low-income countries to agricultural protection in rapidly growing middle- and upper-income countries provides a clear picture of the long-run intertwining of economic and political concerns in the formation of agricultural price policy. Poor countries cannot afford expensive food; their wages would be too high for labor-intensive exports, and their low-income consumers would not be able to afford an adequate diet. As growth proceeds, however, the share of agriculture in an economy shrinks, and rural poverty becomes a pressing issue whenever mobility of resources from rural to urban areas is limited. Most countries have found higher agricultural prices through direct protection to be the only effective mechanism to deal with this pervasive rural poverty. By the time such problems of structural change have become significant, the share of food costs in the budgets of urban workers has usually declined sufficiently that the political costs of higher food prices are manageable. Because the higher agricultural prices tend to be capitalized into land prices, farmers adamantly oppose trade liberalization that would seriously threaten their asset holdings, even if they could easily earn higher incomes in nonfarm occupations. There seems to be something almost inevitable about this dynamic process. The welfare costs are not just in low-income countries that discriminate against their farmers; in the mid-1980s, the costs in Japan, Western Europe, and the United States to consumers and to national budgets are so large that major changes in agricultural price policy are being contemplated.

What would happen if low-income countries raised their farm prices and the OECD countries lowered theirs? No policy maker or donor can reasonably expect an analyst to identify, much less quantify, all of the effects of such changes in price for important food or agricultural commodities. The goal of this essay is much more modest: to provide the analyst with a sense of how markets adjust in the short run and the long run to a variety of government interventions that are designed to alter commodity prices. Price policy analysis must be much more than a checklist of desirable and undesirable attributes of a price intervention or formulas to quantify them. In addition, the need for flexibility in policy design and a capacity for monitoring and evaluating outcomes are stressed.

At some point in a country's development process, basic decisions must be faced about the role markets will be permitted to play in solving the problems of poverty. Markets are fragile instruments for the efficient allocation of resources, but they are tenacious in providing contrary signals to a government whose price policy is badly out of line with available resources. Markets do not solve the problem of poverty, but they strengthen the connections of the poor to a growing economy which provides jobs and incomes—the only hope for them and their children to share in a more abundant future. Policy can strengthen these connections and use them to channel resources and opportunities to those in need. Markets are too important to disappear altogether, but forcing them underground severely limits these opportunities.

References

Alderman, Harold. "Is There 'Curvature' in the Slutsky Matrix?—A Rejoinder." Washington, D.C.: International Food Policy Research Institute, May 1985; processed.

Anderson, Kym, and Yujiro Hayami, with others. *Political Economy of Agricultural Protection: East Asia in International Perspective*. Boston: Allen & Unwin, forthcoming.

Austin, James E. "The Role of Parastatals in Food Policy Implementation." Boston: Harvard Business School, 1984; processed.

Bates, Robert H. *Markets and States in Tropical Africa: The Political Basis of Agricultural Policies*. Berkeley: University of California Press, 1981.

Behrman, Jere R., and Anil B. Deolalikar. "Health and Nutrition." In *Handbook of Development Economics,* ed. Hollis B. Chenery and T. N. Srinivasan. Amsterdam: North-Holland, forthcoming.

Binswanger, Hans P., and Mark R. Rosenzweig. *Contractual Arrangements, Employment and Wages in Rural Labor Markets: A Critical Review*. New York: Agricultural Development Council and International Crops Research Institute for the Semi-Arid Tropics, 1981.

Braverman, Avishay; Choong Yong Ahn; and Jeffrey S. Hammer. "Alternative Agricultural Pricing Policies in the Republic of Korea: Their Implications for Government Deficits, Income Distribution, and Balance of Payments." World Bank Staff Working Paper no. 621. Washington, D.C.: World Bank, 1983.

Brown, Lester R. "Global Food Prospects: Setting the Record Straight." *Challenge Magazine*, November–December 1982.

Clift, F. C. "Review of *Food Policy Analysis*." *Journal of Agricultural Economics*, 35 (September 1984), pp. 425–27.

de Melo, Jaime, and Sherman Robinson. "Trade Adjustment Policies and Income Distribution in Three Archetype Developing Economies." *Journal of Development Economics*, 10 (February 1982), pp. 67–92.

Gillis, S. Malcolm; Dwight H. Perkins; Michael Roemer; and Donald R. Snodgrass. *The Economics of Development*. New York: Norton, 1983.

Hartford, Kathleen. 1983. "Once More with Feeling: A New Stage in Chinese Agricultural Policy." Boston: Harvard Business School, February; processed.

——. 1984. "Managing Socialist Agriculture." Boston: Department of Political Science, University of Massachusetts, Boston; processed.

Hayami, Yujiro, and Vernon Ruttan. *Agricultural Development: An International Perspective*. Baltimore: Johns Hopkins University Press, 1972.

Hazell, Peter B. R., and Ailsa Roëll. *Rural Growth Linkages: Household Expenditure Patterns in Malaysia and Nigeria*. IFPRI Research Report no. 41. Washington: International Food Policy Research Institute, September 1983.

International Food Policy Research Institute. "IFPRI Report 1984." Washington, D.C., 1985.

Johnson, D. Gale. *World Agriculture in Disarray*. New York: Macmillan, 1973.

Johnston, Bruce F., and William C. Clark. *Redesigning Rural Development: A Strategic Perspective*. Baltimore: Johns Hopkins University Press, 1982.

Kirchner, J.; I. Singh; and Lyn Squire. "Agricultural Pricing and Marketing Policies in Malawi." Country Policy Department Discussion Paper. Washington, D.C.: World Bank, 1984.

Koester, Ulrich. *Policy Options for the Grain Economy of the European Community: Implications for Developing Countries*. IFPRI Research Report no. 35. Washington, D.C.: International Food Policy Research Institute, November 1982.

Krishna, Raj. "Measurement of the Direct and Indirect Employment Effects of Agricultural Growth with Technical Change." In *Agricul-*

ture in Development Theory, ed. Lloyd Reynolds. New Haven: Yale University Press, 1975, pp. 297–323.

Kutcher, Gary P.; Alexander Meeraus; and Gerald T. O'Mara. "Agricultural Modeling for Policy Analysis." Preliminary draft. Washington, D.C.: World Bank, February 1985; processed.

Martin, Michael V., and Ray F. Brokken. "The Scarcity Syndrome: A Comment." *American Journal of Agricultural Economics*, 65 (February 1983), pp. 158–59.

Meier, Gerald M., ed., *Pricing Policy for Development Management*. EDI Series in Economic Development. Baltimore: Johns Hopkins University Press for the Economic Development Institute of the World Bank, 1983.

Mellor, John W. "Food Price Policy and Income Distribution in Low-Income Countries." *Economic Development and Cultural Change*, 27, no. 1 (1978), pp. 1–26.

Montias, J. Michael. Review of *Comparative Economic Systems: An Assessment of Knowledge, Theory, and Method*, ed. Andrew Zimbalist, in the Recent Economic Thought series (Boston: Kluwer–Nijhoff, 1984). *Journal of Economic Literature*, 23 (June 1985), pp. 627–28.

Newbery, David M. G., and Joseph E. Stiglitz. *The Theory of Commodity Price Stabilization: A Study in the Economics of Risk*. Oxford: Clarendon Press, 1981.

O'Mara, Gerald, and Vinh Le-Si. "The Supply and Welfare Effects of Rice-Pricing Policy in Thailand." World Bank Staff Working Paper no. 714. Washington, D.C.: World Bank, 1985.

Peck, Anne E. "Futures Markets, Food Imports and Food Security." AGREP Division Working Paper no. 43. Washington, D.C.: World Bank, September 1982.

Pinthong, Chirmsak. "A Price Analysis of Rice Marketing in Thailand." Ph.D. dissertation, Food Research Institute, Stanford University, 1978.

Reich, Michael R.; Yasuo Endo; and C. Peter Timmer. "The Political Economy of Structural Change: Conflict between Japanese and United States Agricultural Policy." Boston: Harvard Business School, January 1986; processed.

Robinson, Sherman. "Multisectoral Models." In *Handbook of Development Economics*, ed. Hollis B. Chenery and T. N. Srinivasan. Amsterdam: North-Holland, forthcoming.

Scandizzo, Pasquale, and Colin Bruce. "Methodologies for Measuring

Agricultural Price Intervention Effects." World Bank Staff Working Paper no. 394. Washington, D.C.: World Bank, July 1980.

Schultz, Theodore W., ed. *Distortions of Agricultural Incentives*. Bloomington: Indiana University Press, 1978.

Schumpeter, Joseph. *Capitalism, Socialism and Democracy*. New York: Harper & Row, 1950.

Simon, Julian L., and William J. Hudson. "Global Food Prospects: Good News." *Challenge Magazine*, November–December 1982.

Srinivasan, T. N. "Malnutrition in Developing Countries: The State of Knowledge of the Extent of Its Prevalence, Its Causes and Its Consequences." New Haven: Department of Economics, Yale University, 1983; processed.

Stiglitz, J. E. "The New Development Economics." In *Handbook of Development Economics*, ed. Hollis B. Chenery and T. N. Srinivasan. Amsterdam: North-Holland, forthcoming.

Streeten, Paul P. "What Price Food?—Agricultural Price Policies in Developing Countries." Washington, D.C.: Economic Development Institute of the World Bank, January 1986; processed.

Taylor, Lance. *Macro Models for Developing Countries*. New York: McGraw-Hill, 1980.

Timmer, C. Peter. 1972. "Employment Aspects of Investment in Rice Marketing in Indonesia." *Food Research Institute Studies in Agricultural Economics, Trade, and Development*, 11, no. 1, pp. 59–88.

——. 1974. "A Model of Rice Marketing Margins in Indonesia." *Food Research Institute Studies*, 13, no. 2, pp. 145–67.

——. 1981. "Is There 'Curvature' in the Slutsky Matrix?" *Review of Economics and Statistics*, 62, no. 3, pp. 395–402.

——. 1983. "The Financial Aspects of Macro Food Policy." In *Research in Domestic and International Agribusiness Management*, vol. 4, ed. Ray A. Goldberg. Greenwich, Conn.: JAI Press, 1983.

——. 1984. "Energy and Structural Change in the Asia-Pacific Region: The Agricultural Sector." In *Energy and Structural Change in the Asia-Pacific Region: Papers and Proceedings of the Thirteenth Pacific Trade and Development Conference*, ed. Romeo M. Bautista and Seiji Naya. Manila: Philippine Institute for Development Studies and the Asian Development Bank, 1984.

——. 1985. "The Role of Price Policy in Increasing Rice Production in Indonesia, 1968–1982." HIID Development Discussion Paper no. 196. Cambridge: Harvard Institute for International Development. Also in *Research in Domestic and International Agribusiness Management,* vol. 6, ed. Ray A. Goldberg. Greenwich, Conn.: JAI Press, 1986.

Timmer, C. Peter; Walter P. Falcon; and Scott R. Pearson. *Food Policy Analysis*. Baltimore: Johns Hopkins University Press for the World Bank, 1983.

Timmer, C. Peter; John W. Thomas; Louis T. Wells; and David Morawetz. *Choice of Technique in Developing Countries: Some Cautionary Tales*. Harvard Studies in International Affairs, no. 37. Cambridge: Harvard Center for International Affairs, 1975.

Tolley, George S.; Vinod Thomas; and Chung Ming Wong. *Agricultural Price Policies and the Developing Countries*. Baltimore: Johns Hopkins University Press for the World Bank, 1982.

Tsakok, Isabelle. "Agricultural Price Policy: A Practitioner's Guide to Partial Equilibrium Analysis." Preliminary draft. Washington, D.C.: World Bank, November 1985; processed.

Waterfield, Charles. "Disaggregating Food Consumption Parameters." *Food Policy,* 10 (November 1985), pp. 337–51.

Wells, Louis T. "Economic Man and Engineering Man: Choice of Technology in a Low-Wage Country." In C. Peter Timmer et al., *Choice of Technique in Developing Countries: Some Cautionary Tales*. Harvard Studies in International Affairs, no. 37. Cambridge: Harvard Center for International Affairs, 1975.

World Bank. *World Development Report 1983*. New York: Oxford University Press for the World Bank, 1983.

Index

Index

Index

Library of Congress Cataloging-in-Publication Data

Timmer, C. Peter.
Getting prices right.

Bibliography: p.
Includes index.
1. Food prices—Government policy—United States.
2. Agricultural prices—Government policy—United States. I. Title
HD9006.T58 1986 338.1'8 86–47649
ISBN 0–8014–1960–3 (alk. paper)